# About the Author

**Charles Liu** is a professor of astrophysics at the City University of New York's College of Staten Island, and an associate with the Hayden Planetarium and Department of Astrophysics at the American Museum of Natural History in New York. His research focuses on colliding galaxies, quasars, starbursts, and the star formation history of the universe. He earned degrees from Harvard University and the University of Arizona, and did postdoctoral research at Kitt Peak National Observatory and at Columbia University. Along with numerous academic research publications, he also writes the astronomy column "Out There" for *Natural History Magazine.* Together with co-authors Neil Tyson and Robert Irion, he received the 2001 American Institute of Physics Science Writing Award for their book *One Universe: At Home in the Cosmos.* He received the 2005 Award for Popular Writing on Solar Physics from the American Astronomical Society. He lives in New Jersey with his wife, daughter, and sons.

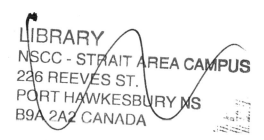

## Also from Visible Ink Press

The Handy Anatomy Answer Book, ISBN 978-1-57859-190-9

The Handy Answer Book for Kids (and Parents), ISBN 978-1-57859-110-7

The Handy Biology Answer Book, ISBN 978-1-57859-150-3

The Handy Geography Answer Book, ISBN 978-1-57859-062-9

The Handy Geology Answer Book, ISBN 978-1-57859-156-5

The Handy History Answer Book, ISBN 978-1-57859-170-1

The Handy Math Answer Book, ISBN 978-1-57859-171-8

The Handy Ocean Answer Book, ISBN 978-1-57859-063-6

The Handy Physics Answer Book, ISBN 978-1-57859-058-2

The Handy Politics Answer Book, ISBN 978-1-57859-139-8

The Handy Presidents Answer Book, ISBN 978-1-57859-167-1

The Handy Religion Answer Book, ISBN 978-1-57859-125-1

The Handy Science Answer Book®, ISBN 978-1-57859-140-4

The Handy Sports Answer Book, ISBN 978-1-57859-075-9

The Handy Supreme Court Answer Book, ISBN 978-1-57859-196-1

The Handy Weather Answer Book®, ISBN 978-0-7876-1034-0

Please visit us at visibleink.com.

# THE
# HANDY
# ASTRONOMY
# ANSWER
# BOOK

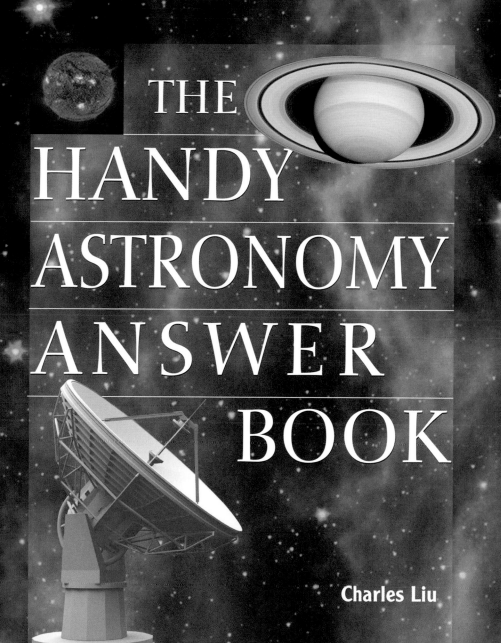

# THE
# HANDY
# ASTRONOMY
# ANSWER
# BOOK

Charles Liu

VISIBLE INK

Detroit

# THE HANDY ASTRONOMY ANSWER BOOK

Visible Ink Press®
43311 Joy Rd., #414
Canton, MI 48187-2075

Visible Ink Press is a registered trademark of Visible Ink Press LLC.

Most Visible Ink Press books are available at special quantity discounts when purchased in bulk by corporations, organizations, or groups. Customized printings, special imprints, messages, and excerpts can be produced to meet your needs. For more information, contact Special Markets Director, Visible Ink Press, www.visibleink.com, or 734-667-3211.

Managing Editor: Kevin S. Hile
Typesetting: Marco Di Vita
Indexer: Lawrence W. Baker
Proofreaders: Sarah Hermsen and Amy Marcaccio Keyzer

ISBN 978-1-57859-193-0

Frontcover images: *Young Stars Emerge from Orion's Head* (NASA/JPL-Caltech/ Laboratorio de Astrofísica Espacial y Física Fundamental); *Saturn's Rings in Visible Light* (NASA and E. Karkoschka, University of Arizona); *Extreme Ultraviolet Imaging Telescope (EIT) image of a huge, handle-shaped prominence* (ESA/NASA/SOHO); and *Radio Frequency Telescope* (iStock)

Backcover image: *Three Moons Cast Shadows on Jupiter* (NASA, ESA, and E. Karkoschka: University of Arizona)

Library of Congress Cataloging-in-Publication Data

Liu, Charles, 1968 Apr. 5-
  The handy astronomy answer book / Charles Liu.
      p. cm.
  Includes index.
  ISBN-13: 978-1-57859-193-0
  ISBN-10: 1-57859-193-7
  1. Astronomy--Miscellanea. I. Title.
  QB52.L58 2008
  520--dc22
                                        2008023254

Printed and bound in China by Imago.

10 9 8 7 6 5 4 3 2 1

# Contents

# Introduction

**W**hy do the stars shine? What happens when you fall into a black hole? What's the Moon made of? Is Pluto a planet or not? Does extraterrestrial life exist? How old is Earth? Can humans live in outer space? What is a quasar? How did the universe begin? How will it end? When it comes to the cosmos, it seems like everyone has a thousand questions.

Well, you're in luck—this book has a thousand answers.

Actually, it contains *more* than a thousand answers to more than a thousand questions about the universe and how it works. These pages contain far more, though, than a mere compilation of facts and figures. Together, these questions and answers tell the story of astronomy—of the cosmos and its contents, and of humanity's efforts throughout history to unlock its secrets and solve its mysteries.

Since the dawn of civilization, people have tried to understand the objects in the heavens—what they are, how they move, and why. At first, it was a total mystery; our ancient ancestors created myths and stories, and ascribed supernatural qualities to the stars and planets. Slowly, they learned that the heavens and its contents were *natural,* not supernatural, and that everyone, not just a privileged few, could understand them. Slowly, the science of astronomy was born.

What is science? It sure isn't a bunch of facts in a big thick book that old folks in lab coats think you should memorize, regurgitate, and forget. Science is a process of asking questions and seeking answers by weighing the facts, making educated guesses, and then testing those guesses with predictions, experiments, and observations. That's what this book is all about: the unquenchable impulse to ask questions and seek answers. You'll read about the questions that were asked, the people who asked them, how they tried to find the answers, and what they discovered in the process. We owe what we know about the universe to the tireless work of those questioners—those men and women who laid the foundation of astronomy, who searched at the frontiers of knowledge.

And that search goes on. In modern times, our species has seen to the edge of the observable universe with ground-based and space-borne telescopes. We have explored distant worlds with robotic spacecraft. We have even started to take our first baby steps into space ourselves. And yet, the more we learn and experience, the more we realize how much we still don't know. This book contains a thousand answers, and that's just a start. May those answers lead you to a thousand more questions; and like those scientific explorers who came before us, may you also experience the joy of discovery as you seek—and find—the answers!

# Acknowledgments

Thank you, Kevin Hile, for being a great editor, and thank you, Roger Jänecke, for being a great publisher. The two of you, more than anyone else, have shepherded this book to its happy completion. To you and those who work with you, I am grateful!

Thank you to Phillis Engelbert and Diane Dupuis, and to everyone who helped them create *The Handy Space Answer Book* back in the late-20th century. Their efforts planted the seed that eventually sprouted this book. Nice work!

And to Amy, Hannah, Allen, and Isaac: thank you! Thank you! Thank you! Thank you! You are the joy and the laughter in my universe.

# ASTRONOMY FUNDAMENTALS

## IMPORTANT DISCIPLINES IN ASTRONOMY

### What is **astronomy**?

Astronomy is the scientific study of the universe and everything in it. This includes, but is not limited to, the study of motion, matter, and energy; the study of planets, moons, asteroids, comets, stars, galaxies, and all the gas and dust between them; and even the study of the universe itself, including its origin, aging processes, and final fate.

### What is **astrophysics**?

Astrophysics is the application of the science of physics to the universe and everything in it. The most important way astronomers gain information about the universe is by gathering and interpreting light energy from other parts of the universe (and even the universe itself). Since physics is the most relevant science in the study of space, time, light, and objects that produce or interact with light, the majority of astronomy today is conducted using physics.

### What is **mechanics**?

Mechanics is the branch of physics that describes the motions of objects in a system. Systems of moving bodies can be very simple, such as Earth and the Moon, or they can be very complicated, such as the Sun, planets, and all the other objects in the solar system put together. Advanced studies of mechanics require complex and detailed mathematical techniques.

## What is **astrochemistry**?

Astrochemistry is the application of the science of chemistry to the universe and everything in it. Modern chemistry—the study of molecules and their interactions—has developed almost exclusively at or near Earth's surface, with its temperature, gravity, and pressure conditions. Its application to the rest of the universe, then, is not quite as direct or ubiquitous as physics is. Even so, astrochemistry is extremely important to cosmic studies: the interactions of chemicals in planetary atmospheres and surfaces is vital to understanding the planets and other bodies in the solar system. Many chemicals have been detected in interstellar gas clouds throughout the Milky Way and other galaxies, including water, carbon monoxide, methane, ammonia, formaldehyde, acetone (which we use in nail polish remover), ethylene glycol (which we use in antifreeze), and even 1, 3-dihydroxyacetone (which is found in sunless tanning lotion).

## What is **astrobiology**?

Astrobiology is the application of the science of biology to the universe and everything in it. This branch of astronomy is very new. The serious use of biology to study the cosmos has blossomed in recent years, however, and has become very important in the field as a whole. With modern astronomical methods and technology, it has become scientifically feasible to search for extraterrestrial life, look for environments where such life could exist, and study how such life could develop.

## What is **cosmology**?

Cosmology is the part of astronomy that specifically examines the origin of the universe. Until the advent of modern astronomy, cosmology was relegated to the domain of religion or abstract philosophy. Today, cosmology is a vibrant part of science and is not limited to gazing out into the cosmos. Current scientific theories have shown that the universe was once far smaller than an atomic nucleus. This means that modern particle physics and high-energy physics, which can be studied on Earth, are absolutely necessary to decipher the mysteries of the very early universe and, ultimately, the very beginning of everything.

## Which of the many **related scientific disciplines** is most **important** to **astronomy**?

Physics is by far the most important and relevant scientific discipline to the study of the universe and everything in it. In fact, in modern times the terms "astronomy" and "astrophysics" are often used interchangeably. That said, all sciences are important to astronomy, and some disciplines that are not very relevant now may someday be extremely vital. For example, if astronomers eventually find extraterrestrial intelligent life, psychology and sociology could become important to the study of the universe as a whole.

# HISTORY OF ASTRONOMY

### When did people first **begin to study** what is now called **astronomy**?

Astronomy is probably the oldest of the natural sciences. Since prehistoric times, humans have looked at the sky and observed the motions of the Sun, Moon, planets, and stars. As humans began to develop the first applied sciences, such as agriculture and architecture, they were already well aware of the celestial objects above them. Astronomy was used by ancient humans to help them keep time and to maximize agricultural production; it probably played an important role in the development of mythology and religion, too.

### What did **early astronomers** use to **measure the universe** before telescopes were invented?

Ancient astronomers, such as Hipparchus (in the second century B.C.E.) and Ptolemy (in the second century C.E.), used instruments such as a sundial, a triquetrum (a sort of triangular ruler), and a plinth (a stone block with an engraved arc) to chart the positions and motions of planets and celestial objects.

By the sixteenth century C.E., complex observational tools had been invented. The famous Danish astronomer Tycho Brahe (1546–1601), for example, crafted many of his own instruments, including a sextant, a quandrant with a radius of six feet (almost two meters), a two-piece arc, an astrolabe, and various armillary spheres.

### What is an **astrolabe**, and how does it work?

An astrolabe is an instrument that can be used by astronomers to observe the relative positions of the stars. It can also be used for timekeeping, navigation, and surveying. The most common type of astronomical astrolabe, called the planispheric astrolabe, was a star map engraved on a round sheet of metal. Around the circumference were markings for hours and minutes. Attached to the metal sheet was an inner ring that moved across the map, representing the horizon, and an outer ring that could be adjusted to account for the apparent rotation of the sky.

To use an astrolabe, observers would hang it from a metal ring attached to the top of the round star map. They could then aim it toward a specific star through a sighting device on the back of the astrolabe, called an adilade. By moving the adilade in the direction of the star, the outer ring

The astrolabe helped mariners navigate the seas for hundreds of years by measuring the positions of the stars. (*iStock*)

would pivot along the circumference of the ring to indicate the time of day or night. The adilade could also be adjusted to measure the observer's latitude and elevation on Earth.

## Who is thought to have **invented** the **astrolabe**?

The ancient Greek mathematician Hypatia of Alexandria (370–415 C.E.) is thought to be the first woman in western civilization to teach and study highly advanced mathematics. During her lifetime, the Museum of Alexandria was a great learning institution with a number of schools, public auditoriums, and what was then the world's greatest library. Hypatia's father, Theon of Alexandria, was the last recorded member of the Museum.

Hypatia was a teacher at one of the Museum's schools, called the Neoplatonic School of Philosophy, and became the school's director in 400 C.E. She was famous for her lively lectures and her many books and articles on mathematics, philosophy, and other subjects. Although very few written records remain, and much information is missing about her life overall, the records suggest that Hypatia invented or helped to invent the astrolabe.

## What is the art of **astrology**?

Astrology is the ancient precursor of the science of astronomy. Ancient people understood that the Sun, Moon, planets, and stars were important parts of the universe, but they could only guess what significance they had or what effects they might cause on human life. Their guesses became a practice in fortune-telling. Astrology was an important part of ancient cultures around the world, but it is not science.

## What did **ancient Middle Eastern cultures** know about **astronomy**?

The Mesopotamian cultures (Sumerians, Babylonians, Assyrians, and Chaldeans) were very knowledgable about the motions of the Sun, Moon, planets, and stars. They mapped the 12 constellations of the zodiac. Their towering temples, called ziggurats, may have been used as astronomical observatories. Arab astronomers built great observatories throughout the Islamic empires of a thousand years ago, and we still use Arabic names for many of the best-known stars in the sky.

The ruins of Mexico's Chichen Itza, where Anasazi astronomers observed the skies and accurately calculated lunar cycles, equinoxes, and solstices. (*iStock*)

## What did **ancient American cultures** know about **astronomy**?

Ancient American cultures were very knowledgeable about astronomy, including lunar phases, eclipses, and planetary motions. Almost all of the many temples and pyramids of the Inca, Mayan, and other Meso-American cultures are aligned and decorated with the motions of planets and celestial objects.

For example, at Chichen Itza in southern Mexico, on the days of the vernal equinox (March 21) and autumnal equinox (September 21), shadows cast by the Sun create the vision of a huge snake-god slithering up the sides of the Pyramid of Kukulcan, which was built more than one thousand years ago. Farther north, among the Anasazi ruins of Chaco Canyon, New Mexico, the work of ancient Native American astronomers survives in the famous "sun dagger" petroglyphs, which appear to mark the solstices, equinoxes, and even the 18.67-year lunar cycle.

## What is the *Dresden Codex*, and what does it say about **Mayan astronomy**?

There are three well-known records from what is believed to have been an extensive Mayan library, dating back perhaps one thousand years to the height of the Mayan civilization. One of these books is called the *Dresden Codex* because it was discovered in the late 1800s in the archives of a library in Dresden, Germany. It includes observations of the motions of the Moon and Venus, and predictions of the times at which lunar eclipses would occur.

## What is Stonehenge?

Stonehenge is one of the world's most famous ancient astronomical sites. This assembly of boulders, pits, and ditches is located in southwestern England, about eight miles (13 kilometers) away from the town of Salisbury. Stonehenge was built and rebuilt during a period from about 3100 B.C.E. to 1100 B.C.E. by ancient Welsh and British nature-worshipping priests called druids.

Archaeologists think Stonehenge had astronomical significance. It was certainly built with astronomical phenomena in mind. One pillar, called the Heel Stone, appears to be near the spot where sunlight first strikes on the summer solstice. Thus, Stonehenge may have served as a sort of calendar. Other evidence suggests that Stonehenge may have been used as a predictor of lunar eclipses.

Perhaps the most remarkable section of the *Dresden Codex* is a complete record of the orbit of Venus around the Sun. Mayan astronomers had correctly calculated that it takes Venus 584 days to complete its orbit. They arrived at this figure by counting the number of days that Venus first appeared in the sky in the morning, the days when it first appeared in the evening, and the days that it was blocked from view because it was on the opposite side of the Sun. The Mayans then marked the beginning and ending of the cycle with the heliacal rising, the day on which Venus rises at the same time as the Sun.

## What did **ancient East Asian cultures** know about **astronomy**?

Some of the world's earliest astronomical observations were made by the ancient Chinese. Perhaps as early as 1500 B.C.E., Chinese astronomers created the first rough charts of space. In 613 B.C.E., they described the sighting of a comet. Within a few centuries after that, Chinese astronomers were keeping track of all the eclipses, sunspots, novae, meteors, and celestial and sky phenomena they observed.

Chinese astronomers made numerous contributions to the field of astronomy. They studied, for instance, the question of Earth's motion and created one of the earliest known calendars. By the fourth century B.C.E., Chinese astronomers had produced a number of star charts, which depicted the sky as a hemisphere—a perfectly logical strategy, since we can only see half the sky at any one time. Three centuries after that, Chinese astronomers began to regard space as an entire sphere, showing they were aware of Earth's spherical shape, as well as of Earth's rotation around its polar axis. They created an early map of the celestial sphere on which they placed stars in relation to the Sun and to the North Star.

Chinese astronomers were the first to observe the Sun; they protected their eyes by looking through tinted crystal or jade. The Sung Dynasty, which began in

England's ancient Stonehenge may have served as a type of astronomical calendar used by the druids. (*iStock*)

960 C.E., was a period of great astronomical study and discovery in China. Around this time, the first astronomical clock was built and mathematics was introduced into Chinese astronomy.

## What did **ancient African cultures** know about **astronomy**?

The ancient Egyptians built their pyramids and other great monuments with a clear understanding of the rhythms of rising and setting celestial objects. The Egyptians established the 365-day solar year calendar as early as 3000 B.C.E. They established the 24-hour day, based on nightly observations of a series of 36 stars (called decan stars). At midsummer, when 12 decans were visible, the night sky was divided into 12 equal parts—the equivalent to hours on modern clocks. The brightest star in the night sky, Sirius the "Dog Star," rose at the same time as the Sun during the Egyptian midsummer; this is the origin of the term "dog days of summer."

## What did **other ancient cultures** around the world know about **astronomy**?

A knowledge of the night sky seems to be a common thread among all the major cultures and societies of the ancient world. Polynesian cultures, for instance, used the Pleiades (the cluster of stars also known as "The Seven Sisters") to navigate around the Pacific Ocean. Australian aborigine cultures, south Asian cultures, Inuit cultures, and northern European cultures all had their own sets of myths and legends about the motions of the Sun and the Moon, as well as their own maps of the stars and of constellations.

## What contributions did **ancient Greek astronomers** make to the science of astronomy?

The contributions of ancient Greek astronomers are numerous. Many of them were also pioneers in mathematics and the origins of scientific inquiry. Some notable examples include Eratosthenes (c. 275–195 B.C.E.), who first made a mathematical measurement of the size of Earth; Aristarchus (c. 310–230 B.C.E.), who first hypothesized that Earth moved around the Sun; Hipparchus (c. 190–120 B.C.E.), who made accurate star charts and calculated the geometry of the sky; and Ptolemy (c. 85–165 C.E.), whose model of the solar system dominated the thinking of Western civilization for more than a thousand years.

## What is the **Ptolemaic model** of the **solar system**?

About 140 C.E. the ancient Greek astronomer Claudius Ptolemy, who lived and worked in Alexandria, Egypt, published a 13-volume treatise on mathematics and astronomy called *Megale mathmatike systaxis* ("The Great Mathematical Compilation"), which is better known today as *The Almagest*. In this work, Ptolemy built upon—and in some cases, probably reprised—the work of many predecessors, such as Euclid, Aristotle, and Hipparchus. He described a model of the cosmos, including the solar system, that became the astronomical dogma in Western civilization for more than one thousand years.

According to the Ptolemaic model, Earth stands at the center of the universe, and is orbited by the Moon, the Sun, Mercury, Venus, Mars, Jupiter, and Saturn. The stars in the sky are all positioned on a celestial sphere surrounding these other objects at a fixed distance from Earth. The planets follow circular orbits, with extra "additions" on their orbital paths known as epicycles, which explain their occasional retrograde motion through the sky. Ptolemy also cataloged more than one thousand stars in the night sky. Although the Ptolemaic model of the solar system was proven wrong by Galileo, Kepler, Newton, and other great scientists starting in the seventeenth century, it was very important for the development of astronomy as a modern science.

# MEDIEVAL AND RENAISSANCE ASTRONOMY

### What influence did the **Catholic Church** have on **astronomy** in **medieval Europe**?

Most historians agree that the immense power of the Catholic Church during the Middle Ages stifled astronomical study in Europe during that time. One tenet of Catholic dogma stated that space is eternal and unchanging; so, for example, when people observed a supernova in 1054 C.E. its occurrence was recorded in other parts of the world but not in Europe. Another part of Church dogma erroneously declared that the Sun, Moon, and planets moved around Earth. By the 1500s, a thousand years after the fall of Rome, the Catholic Church finally began to contribute again to the science of astronomy, such as with the development of an accurate calendar.

### Who first began the **challenge** to the **geocentric model** of the **solar system**?

Polish mathematician and astronomer Nicholas Copernicus (1473–1543; in Polish, Mikolaj Kopernik) suggested in 1507 that the Sun was at the center of the solar system, not Earth. His "heliocentric" model had been proposed by the ancient Greek astronomer Aristarchus around 260 B.C.E., but this theory did not survive past ancient times. Copernicus, therefore, was the first European after Roman times to challenge the geocentric model.

### How did **Copernicus present** the **heliocentric model** of the solar system?

Copernicus wrote his ideas in *De Revolutionibus Orbium Coelestium,* which was published just before his death in 1543. In this work, Copernicus presented a heliocentric model of the solar system in which Mercury, Venus, Earth, Mars, Jupiter, and Saturn moved around the Sun in concentric circles.

### How did the **heliocentric model** of the solar system advance **after the death** of **Copernicus**?

Unfortunately, *De Revolutionibus Orbium Coelestium* was placed on the Catholic Church's list of banned books in 1616, where it remained until 1835. Before it was banned, word of the heliocentric model nonetheless spread among astronomers and scholars. Eventually, Galileo

Nicholas Copernicus. *(Library of Congress)*

9

Galileo Galilei. *(Library of Congress)*

Galilei (1564–1642) used astronomical observations to prove that the heliocentric model was the correct model of the solar system; Johannes Kepler (1571–1630) formulated the laws of planetary motion that described the behavior of planets in the heliocentric model; and Isaac Newton (1642–1727) formulated the Laws of Motion and the Law of Gravity, which explained why the heliocentric model works.

## Who was **Galileo Galilei**?

Italian scholar Galileo Galilei (1564–1642) is considered by many historians to be the first modern scientist. One of the last great Italian Renaissance men, Galileo was born in Florence and spent a good deal of his professional life there and in nearby Padova. He explored the natural world through observations and experiments; wrote eloquently about science and numerous other philosophical topics; and rebelled against an established authority structure that did not wish to acknowledge the implications of his discoveries. Galileo's work paved the way for the study and discovery of the laws of nature and theories of science.

### How did **Galileo contribute** to our **understanding** of the **universe**?

Galileo was the first person to use a telescope to study space. Even though his telescope was weak by modern standards, he was able to observe amazing cosmic sights, including the phases of Venus, mountains on the Moon, stars in the Milky Way, and four moons orbiting Jupiter. In 1609 he published his discoveries in *The Starry Messenger,* which created a tremendous stir of excitement and controversy.

Galileo's observations and experiments of terrestrial pheonomena were equally important in changing human understanding of the physical laws of the cosmos. According to one famous story, he dropped metal balls of two different masses from the top of the Leaning Tower of Pisa. They landed on the ground at the same time, showing that an object's mass has no effect on its speed as it falls to Earth. Through his works *A Dialogue Concerning the Two Chief World Systems* and *Discourse on Two New Sciences,* Galileo described the basics of how objects move both on Earth and in the heavens. These works led to the origins of physics, as articulated by Isaac Newton and others who followed him.

### What happened between **Galileo** and the **Catholic Church**?

Galileo's support of the heliocentric model was considered a heretical viewpoint in Italy at the time. The Catholic Church, through its Inquisition, threatened to torture or even kill him if he did not recant his writings. Ultimately, Galileo did recant his discoveries and lived under house arrest for the last decade of his life. It is said that, in a private moment after his public recantation, he stamped his foot on the ground and said, "Eppe si muove" ("Nevertheless, it moves.")

## Who was **Tycho Brahe**?

Tycho Brahe (1546–1601), despite being a Danish nobleman, turned to astronomy rather than politics. Granted the island of Hven in 1576 by King Frederick II, he established Uraniborg, an observatory containing large, accurate instruments. Uraniborg was the most technologically advanced facility of its type ever built. Brahe's measurements of planetary motions, therefore, were more precise than any that had been previously obtained. This facility and these measurements helped Brahe's protégé, Johannes Kepler, determine the elliptical nature of the motion of planets around the Sun.

## Who was **Johannes Kepler**?

German astronomer Johannes Kepler (1571–1630) was very interested in the mathematical and mystical relationships between objects in the solar system and geometric forms such as spheres and cubes. In 1596, before working as an astronomer, Kepler published *Mysterium Cosmographicum,* which explored some of these ideas. Later, working with Danish astronomer Tycho Brahe and his data, Kepler helped establish the basic rules describing the motions of objects moving around the Sun.

## How did **Johannes Kepler** **contribute** to our **understanding** of the **universe**?

Kepler worked with Tycho Brahe until Brahe's death in 1601. He succeeded Brahe as the official imperial mathematician to the Holy Roman Emperor. This position gave him access to all of Brahe's data, including his detailed observations of Mars. He used that data to fit the orbital path of Mars using an ellipse rather than a circle. In 1604, he observed and studied a supernova, which he thought was a "new star." At its peak, the supernova was nearly as bright as the planet Venus; today, it is known as Kepler's supernova. Using a telescope he constructed, he verified Galileo's discovery of Jupiter's moons, calling them satellites. Later in his career, Kepler published a book on comets and a cata-

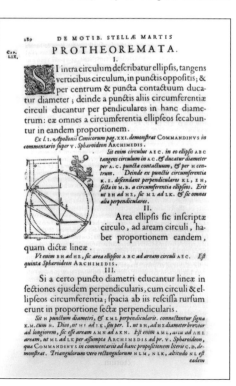

A diagram by Johannes Kepler from his 1609 work *Astronomia nova,* depicting Mars orbiting the sun to illustrate two of his laws of planetary motion. *(Library of Congress)*

log of the motions of the planets called *The Rudolphine Tables* that was used by astronomers throughout the following century. Kepler is perhaps most famous for developing his three laws of planetary motion.

### What is **Kepler's First Law** of **planetary motion**?

According to Kepler's First Law, planets, comets, and other solar system objects travel on an elliptical path with the Sun at one focus point. The effect can be subtle or profound; Earth's orbit, for example, is very nearly circular, whereas the orbit of Pluto is noticeably oblong, and the orbits of most comets are highly elongated.

### What is **Kepler's Second Law** of **planetary motion**?

According to Kepler's Second Law, planetary orbits sweep out equal times in equal areas. This means that a planet will move faster when it is closer to the Sun, and slower when it is farther away. Future scientists such as Isaac Newton showed that the Second Law is true because of an important property of moving systems called the conservation of angular momentum.

### What is **Kepler's Third Law** of **planetary motion**?

According to Kepler's Third Law, the cube of the orbital distance between a planet and the Sun is directly proportional to the square of the planet's orbital period. Kepler discovered this law in 1619, ten years after the publication of his first two laws of planetary motion. It is possible to use this third law to calculate the distance between the Sun and any planet, comet, or asteroid in the solar system, just by measuring the object's orbital period.

Christian Huygens. *(Library of Congress)*

### Who was **Christian Huygens**?

Dutch astronomer, physicist, and mathematician Christian Huygens (1629–1695) is one of the most important figures in the history of science. He was a key transitional scientist between Galileo and Newton. His work was crucial to the development of the modern sciences of mechanics, physics, and astronomy. Huygens helped develop the Law of Conservation of Momentum, invented the pendulum clock, and was the first person to describe a wave theory of light. He designed and built the clearest lenses and most powerful telescopes of his time. Using these tools, he was the first person to identify Saturn's ring system, and he discovered Saturn's largest moon, Titan.

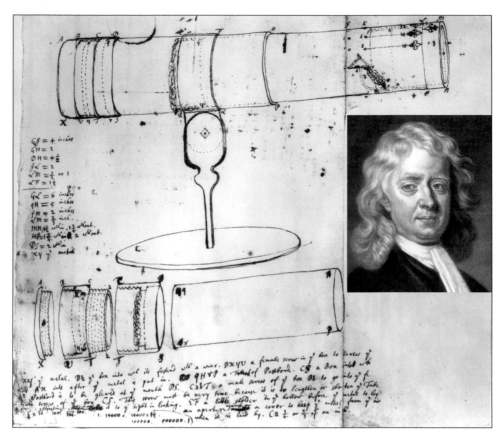

Isaac Newton (inset) and an illustration he drew of the telescope he invented. *(Library of Congress)*

## Who was **Isaac Newton**?

English mathematician, physicist, and astronomer Isaac Newton (1642–1727) is considered to be one of the greatest geniuses who ever lived. He had to leave Cambridge University in 1665 and work on his family farm when the university was closed due to an outbreak of bubonic plague. During the next two years, he made a series of remarkable advances in mathematics and science, including the calculus and his laws of motion and universal gravitation. Newton returned to Cambridge University in 1667, and eventually assumed the position of the Lucasian Professorship of Mathematics. While there, he made fundamental discoveries about optics, invented a new kind of telescope, and published his greatest work, the *Principia,* in 1687, with the encouragement and financial backing of his acquaintance, the astronomer Edmund Halley.

In his later career, he won a seat in the British Parliament and was appointed Master of the Royal Mint. He invented the idea of putting ridges around the edges of coins so people could not shave the coins and keep the precious metals for themselves. The Queen of England knighted him in 1705, the first scientist to be given such an honor. He was also elected head of the Royal Society, the most significant

13

academic body in the world at that time. Sir Isaac Newton died on March 31, 1727, in London, England.

## How did **Newton contribute** to our **understanding** of the **universe**?

In his work *Philosophiae Naturalis Principia Mathematica* ("Mathematical Principles of Natural Philosophy"), or just *Principia,* Newton articulated the law of universal gravitation and his three laws of motion. He also described, in other works, major advances in many areas of knowledge. In optics, he showed that sunlight is really a combination of many colors; in mathematics, he developed new methods that form much of the modern foundation of mathematics, including the calculus, which was also developed by German philosopher and mathematician Gottfried Wilhelm von Leibniz. In cosmology, he supplied a theoretical framework that modern astronomers used to calculate the density of an expanding universe; while in astronomy, he invented a kind of telescope that uses mirrors rather than lenses. It is the basis of all major astronomical research telescopes built today.

## What is **Newton's First Law of Motion**?

According to Newton's First Law, "Every body continues in its state of rest, or of uniform motion in a right line, unless it is compelled to change that state by forces impressed upon it." This is also known as the law of inertia; it simply means that an object tends to stay still, or stay in motion in a straight line, unless it is pushed or pulled. This law is an expression in words of a fundamental property of motion called the conservation of linear momentum. Mathematically, the momentum of an object is its mass multiplied by its velocity.

## What is **Newton's Second Law of Motion**?

According to Newton's Second Law, "The change of motion is proportional to the motive force impressed and is made in the direction of the right line in which that force is impressed." This is also known as the law of force, and it defines force as the change in the amount of motion, or momentum, of an object. Mathematically, the force of an object is its mass multiplied by its acceleration.

### What is **Newton's Third Law of Motion**?

According to Newton's Third Law, "To every action there is always opposed an equal reaction: or the mutual actions of two bodies upon each other are always equal and directed to contrary parts." That means that to exert a force on an object the thing doing the exerting must experience a force of equal strength in exactly the opposite direction. This law explains, for example, why an ice skater goes backward when she pushes another skater forward.

### What is **Newton's Law of Gravity**?

According to Newton's Law of Universal Gravitation, every object in the universe exerts a pulling force on every other object; this force between any two objects is directly proportional to the masses of the two objects multiplied with one another, and it is inversely proportional to the square of the distance between the two objects. In other words, gravity follows what is known as the "inverse square law": a mathematical relationship that governs both the strength of gravity and the propagation of light in space.

### What was the **importance** of Newton's **Law of Gravity** to **astronomy**?

Newton's law of universal gravitation shows that the objects in the solar system move according to a mathematically predictable set of rules. It shows scientifically why Kepler's three laws of orbital motion are true, and it allows astronomers to predict the locations and motions of celestial objects. When Edmund Halley, for example, used the law to predict the 76-year orbital period of a well-known comet—a prediction confirmed after Halley's death—it marked a milestone in astronomy: the final transformation from superstition and ignorance to science and knowledge.

# EIGHTEENTH- AND NINETEENTH-CENTURY ADVANCES

### What significant **scientific advances** occurred in the **1700s** that most **advanced astronomy**?

In the 1700s, the study of mathematics beyond the calculus first established by Leibniz and Newton led to the development of the branch of physics called mechanics. Scientists began to understand the nature of electricity through experiments in laboratories and with lightning. Opticians began to develop telescopes that could let astronomers observe objects invisible to the unaided eye. And using those telescopes, astronomers began to take systematic surveys of the sky, making detailed sky catalogs.

## Who was **Pierre-Simon de Laplace** and what did he contribute to **mechanics**?

Pierre-Simon de Laplace. *(Library of Congress)*

French mathematician and astronomer Pierre-Simon de Laplace (1749–1827) made a number of key contributions to mathematics, astronomy, and other sciences. Together with chemist Antoine-Laurent Lavoisier, Laplace helped develop our understanding of the interrelationship of chemical reactions and heat. In physics, Laplace applied the calculus, recently invented by Isaac Newton and Gottfried Wilhelm von Leibniz, to calculate the forces acting between particles of matter, light, heat, and electricity. Laplace and his colleagues created systems of equations that explained the refraction of light, the conduction of heat, the flexibility of solid objects, and the distribution of electricity on conductors.

In astronomy, Laplace was primarily interested in the movements of the objects in the solar system and their complex gravitational interactions. He published his results over many years in a multi-volume book called *Traite de Mechanique Celeste* ("Celestial Mechanics"). The first volume of *Celestial Mechanics* was published in 1799. Laplace also developed a nebular theory of the formation of the Sun and our solar system, and, along with his colleague John Michel, he introduced the idea of a "dark star," which later came to be called a black hole. Because of his brilliance, and since his work expanded on the gravitational theories of Isaac Newton, Laplace earned the nickname "The French Newton."

## Who was **Joseph-Louis Lagrange** and what did he contribute to **mechanics**?

Joseph-Louis Lagrange (1736–1813) was an Italian mathematician who developed some of the most important theories of mechanics, both regarding Earth and the universe. Generally remembered as a French scientist because he spent the last part of his career in Paris, his analysis of the wobble of the Moon about its axis of rotation won him an award from the Paris Academy of Sciences in 1764. Lagrange also worked on an overall description of the way that forces act on groups of moving and stationary objects, a project that Galileo Galilei and Isaac Newton had begun years before. He eventually succeeded in devising several key general mathematical tools to analyze such forces. These were published in a 1788 work called *Mechanique Analytique* ("Analytical Mechanics"). Lagrange went on to explore the interaction between objects in the solar system as a complex system of objects; he discovered what are called Lagrange points: places around and between two gravitationally bound bodies where a third object could stay stationary relative to the other two. This proves useful today for placing satellites in space.

In 1793, Lagrange was appointed to a commission on weights and measures, and helped create the modern metric system. He spent his final working years trying to develop new mathematical systems of calculus.

### Who was **Leonhard Euler** and what did he contribute to **mechanics**?

The Swiss mathematician Leonhard Euler (1707–1783) was probably the most prolific mathematician in recorded history. He helped unify the systems of calculus first created independently by Leibniz and Newton. He made key contributions to geometry, number theory, real and complex analysis, and many other areas of mathematics. In 1736, Euler published a major work in mechanics, appropriately called *Mechanica,* which introduced methods of mathematical analysis to solve complex problems. Later, he published another work on hydrostatics and rigid bodies, and he did tremendous work on celestial mechanics and the mechanics of fluids. He even published a 775-page work just on the motion of the Moon.

### Who was **Adrien-Marie Legendre** and what did he contribute to **mechanics**?

The French mathematician Adrien-Marie Legendre (1752–1833) taught at the French military academy with Pierre-Simon de Laplace, starting in 1775. In 1782 he won a prize for the best research project on the speed, path, and flight dynamics of cannonballs moving through the air. Elected to the French Academy of Sciences the next year, he combined his research on abstract mathematics with important work on celestial mechanics. In 1794, Legendre wrote a geometry textbook that was the definitive work in the field for nearly a century. In 1806, he published *Nouvelles methods pour la determination des orbits des cometes* ("New Methods for the Determination of the Orbits of Comets"). Here he introduced a technique for finding the equation of a mathematical curve using imperfect data. Legendre is best known today for his work on elliptical functions and for inventing a class of functions called Legendre polynomials, which are valuable tools for studying harmonic vibrations and for finding mathematical curves that fit large series of data points.

### Who created the **New General Catalog**?

The German-English astronomer Caroline Herschel (1750–1848) and her nephew John Herschel (1792–1871) created the New General Catalog (NGC), a list of thousands of astronomical objects that represent most of the best-known gaseous nebulae, star clusters, and galaxies in the night sky.

### What significant **scientific advances** occurred in the **1800s** that most **advanced astronomy**?

In the 1800s, the scientific understanding of electricity and magnetism grew to the point where it was possible to generate controlled amounts of energy from electric-

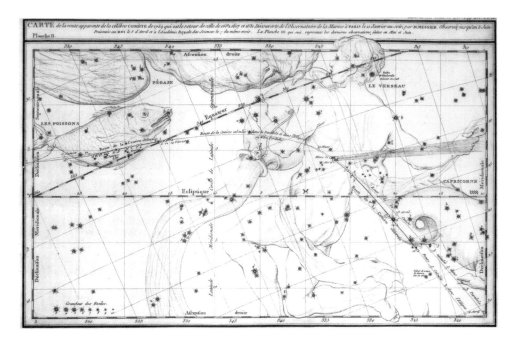

An illustration by Charles Messier from his famous catalog describing the path of Halley's comet. *(Library of Congress)*

ity using generators, and to transport that electricity across large distances. This research led to the understanding of electromagnetism as a force, the transference of electromagnetic energy in the form of waves, and the manifestation of those waves as the electromagnetic spectrum.

Scientists also made major advances in understanding the concept of energy and how it can be manifested in many different forms such as motion, heat, and light. The science of thermodynamics—the study of heat energy and how it is transferred—and its closely related branch of physics, statistical mechanics, were born. These discoveries and their technological applications transformed all of human society: the steam engine, the electric light, and the Industrial Revolution are just a few examples of their impact. Their impact on astronomy was equally significant.

### Who was **James Clerk Maxwell** and what did he contribute to **physics**?

The Scottish scientist and mathematician James Clerk Maxwell (1831–1879) made huge discoveries in a number of areas. In 1861, he produced the first color photograph. He studied the rings of Saturn, theorizing that they were composed of millions of tiny particles rather than solid or liquid structures. He also helped develop the kinetic theory of gases; and his theory of electromagnetism tied together the relationship between electricity and magnetism. Between 1864 and 1873, Maxwell showed that light is actually electromagnetic radiation. A set of four equations known as Maxwell's Equations shows the most basic mathematical and physical relationships between electricity, magnetism, and light.

## Who created the Messier catalog?

French astronomer Charles Messier (1730–1817) was a famed discoverer of comets. Discovering comets with a telescope was a very difficult task at the time, and successes brought the discoverer great fame and prestige. Messier discovered more than a dozen comets. He also discovered a number of objects in the night sky that looked like they might be comets but were not.

Around 1770 Messier started to publish catalogs of the objects he had found with his telescope. Other astronomers later added to the 45 objects originally listed in the Messier catalog, as it became known. The modern version of the Messier catalog contains 110 objects, many of which are the most beautiful and interesting astronomical objects in the night sky.

## Who was **Rudolf Heinrich Hertz** and what did he contribute to **physics**?

The German physicist Rudolf Heinrich Hertz (1857–1894) was a genius in both science and languages (he learned Arabic and Sanskrit as a youth). Aside from his work on electrodynamics, he conducted research on meteorology and contact mechanics (what happens to objects when they are put against one another).

Hertz proved the existence of electromagnetic waves in 1888. Although visible light was known to be electromagnetic in origin, Hertz produced electromagnetic waves not visible to the human eye—radio waves—using a wire connected to an induction coil, then detecting them using a loop of wire and a spark gap. Hertz built upon Maxwell's work, and in 1892 rewrote Maxwell's equations of electrodynamics in the elegant, symmetric form that is most commonly used today. Today, his work is the scientific foundation of all wireless communications, and the unit of electromagnetic frequency is named in his honor.

## Who was **James Joule** and what did he contribute to **physics**?

The English physicist James Prescott Joule (1818–1889) was the son of a wealthy brewer. Although many of his discoveries were not widely accepted for many years, by the end of his career he had made significant contributions to the understanding of how different forms of energy (such as electrical, kinetic, and heat energy) are related. Today, along with the German physician and scientist Julius Robert von Mayer (1814–1878), Joule is credited with figuring out the mathematical conversion factor between heat and kinetic energy. The physical unit for kinetic energy is called the Joule in his honor (one Joule is equal to 0.239 calories).

## Who was **Lord Kelvin** and what did he contribute to **physics**?

The British scientist William Thomson, Lord Kelvin (1824–1907), was a brilliant scientist. The son of an engineering professor, Kelvin published more than 600 scientific articles in his career on a wide variety of topics in the physical sciences. As

19

an applied scientist, he invented a number of scientific instruments; one of them, the mirror-galvanometer, was used in the first successful trans-Atlantic underwater telegraph cable, which ran from Ireland to Newfoundland. His success in applied science earned him fame, wealth, and a noble title: Baron Kelvin of Largs.

In theoretical science, Kelvin was a pioneer in tying together ideas about electricity and magnetism, heat and light, and thermal and gravitational energy. He worked with James Joule (1818–1889) in formulating the first law of thermodynamics, and concluded that there exists an "absolute zero" temperature (the lowest possible temperature in the universe). Today, the temperature scale based on absolute zero is called the Kelvin scale in his honor.

# MATTER AND ENERGY

### What is **energy**?

Energy is that which makes things happen in the universe. It is that which is exchanged between any two particles in order for those particles to change—their motion, their properties, or anything else—in any way. Energy is everywhere around us; it takes so many different forms that it is hard to pin down. Heat is energy; light is energy; everything that moves carries kinetic energy. Even matter itself can be converted into energy, and vice versa.

### What is **matter**?

Matter, the stuff out of which every object in the universe is made, is everything in the universe that has mass. Mass is a quality that is hard to describe. Very roughly, it is the "drag" through spacetime that an object experiences. An object with more mass will move more slowly through spacetime than an object with less mass, if both have the same amount of either momentum or kinetic energy.

### What is the significance of the formula $E = mc^2$?

$E = mc^2$ was discovered by Albert Einstein in 1905. It is a major result of his Special Theory of Relativity, which describes the relationship between how objects and electromagnetic radiation move through space and how they move through time. It means that the amount of energy in a piece of matter is equal to the mass of that piece of matter multiplied by the speed of light squared. This is a huge amount of energy, by the way, for even a tiny amount of matter; the energy contained in a penny far exceeds the explosive power of the atomic bombs detonated in 1945 over Hiroshima and Nagasaki *combined*.

### What is **light**?

Light is a kind of energy. It travels as waves and is carried as particles called photons. Generally speaking, light is electromagnetic radiation. (Radiation carried by massive particles, however, such as alpha rays and beta rays, is not light.) What is interesting about light is that it can be treated as both a stream of particles and as a wave of radiation. The double nature of light—known as "wave-particle duality"—is a cornerstone of the branch of physics called quantum mechanics.

## What are **photons**?

Photons are special subatomic particles that contain and carry energy but have no mass. Photons, in fact, can be imagined as particles of light. Photons are produced or destroyed whenever electromagnetic force is transferred from one place to another.

## What are **electromagnetic waves**?

Electromagnetic waves are electromagnetic radiation, which is light. Usually, on Earth, humans think of light just as the kind of radiation that our eyes can detect.

## What kinds of **electromagnetic radiation** are there?

There are seven general kinds of electromagnetic radiation: gamma rays, X rays, ultraviolet, visible, infrared, microwaves, and radio waves. Gamma rays, X rays, and ultraviolet rays have shorter wavelengths than those of visible light; infrared waves, microwaves, and radio waves have wavelengths longer than those of visible light.

## What is the **speed** of an **electromagnetic wave**?

The speed of light is the same as the speed of an electromagnetic wave because they are the same thing.

## What is the **speed of light**?

Light travels through a vacuum at almost exactly 186,282.4 miles (299,792.5 kilometers) per second, or 670 million miles (1.078 billion kilometers) per hour, or 5.8 trillion miles (9.2 trillion kilometers) per year! A beam of light can go from New York to Tokyo in less than one-tenth of a second, and from Earth to the Moon in less than 1.3 seconds.

## How have scientists **measured** the **speed of light**?

In the late 1500s, Galileo Galilei documented an experiment in which he tried to meas-ure the speed of light by using lanterns on two distant hilltops. He was only able to say that it was much faster than he could measure. In 1675 Danish astronomer Olaus Roe-mer (1644–1710) used eclipses of the moons of Jupiter to measure the speed of light to be 141,000 miles per second, or about 76 percent of the modern value. Roemer came fairly close, but more importantly he showed that the speed of light was not infinite. That discovery had important implications on all of physics and astronomy.

In the mid-1700s, English astronomer James Bradley (1693–1762) noticed that some stars appeared to be moving because Earth was actually moving toward or away from the starlight that was coming toward us. Using this phenomenon, called the aberration of starlight, Bradley was able to measure the speed of light to an accuracy of less than one percent error: 185,000 miles per second. In the 1800s, the French scientist Jean-Bernard León Foucault (1819–1868) used a laboratory setup of two mirrors, one rotating and one unmoving, to measure the speed of light. As the spinning mirror reflected a light beam back and forth from the stationary one, it reflected the beam back at different angles. By using geometry, Foucault deter-mined the speed of light to be just over 186,000 miles per second.

In 1926 American physicist Albert Abraham Michelson (1852–1931) repeated Foucault's experiment on a much larger scale. Using mirrors positioned 22 miles apart on two mountains in California, he calculated light speed to be 186,271 miles per second.

## Does **light** ever **change speed**?

Yes, light can change speed and direction when it goes through different materials. All materials that transmit light have a property called index of refraction. The index of refraction is 1 for a perfect vacuum, 1.0003 for air, 1.33 for water, about 1.5 for various kinds of glass, and 2.42 for diamond. Light travels more slowly through higher index-of-refraction materials than through lower ones.

## What does it mean when we say the **speed of light** is **constant**?

Saying that the speed of light is constant means that any observer watching any par-ticular beam of light will measure that beam to be moving at the same speed. It does not matter whether the observer is moving toward, away, or not at all relative to the

## How did the Michelson-Morley experiment work?

The Michelson-Morley experiment was based on a special experimental technique called interferometry. A beam of light was sent to a silvered mirror set at an angle; some of the light would travel through the mirror, and the rest would bounce off the mirror. Each partial beam of light would then bounce off other mirrors, recombine at the silvered mirror, and then return to the original location of the light source. If the partial beams of light were altered during their travel, the recombined light beam would show a measurable interference pattern.

Since the two light paths had different directions of travel, Michelson and Morley hypothesized that they would interact differently with the luminiferous ether, and thus produce an interference pattern. To their surprise, the recombined beam showed no measurable interference. This null result implied that, despite traveling in different directions for a time, the speed of both beams had remained exactly the same. If any sort of luminiferous ether existed in the universe, this result would not be possible.

beam; it also does not matter how fast the observer is moving. In other words, light does not have the usual kind of relativity when it comes to the relative observed speeds of objects; it follows a special theory of relativity, which was articulated by Albert Einstein in 1905.

## Who first obtained scientific **evidence** that the **speed of light** is **constant**?

Polish-born American physicist Albert Abraham Michelson (1852–1931) and American chemist Edward Williams Morley (1838–1923) conducted an experiment to test the way light travels through the universe. In the late 1800s, scientists thought that light waves traveled through a special substance called "luminiferous ether" in much the same way that ocean waves move through water. The Michelson-Morley experiment was designed to test the properties of the luminiferous ether. The result, however, was not at all what they or other scientists expected. Instead, that experiment showed that the luminiferous ether does not exist and that the speed of light is constant.

## Who studied the **results** of the **Michelson-Morley experiment**?

After the results of the Michelson-Morley experiments were confirmed, many of the leading physicists of the day carefully pondered their implications. The Irish mathematical physicist George Francis Fitzgerald (1851–1901), the Dutch physicist Henrik Antoon Lorentz (1853–1928), and the French mathematician and physicist Jules-Henri Poincaré (1854–1912) were three of the scientists particularly interested in explaining why this result came about. They were able to show that a specific

23

mathematical relationship exists between the length of an object and speed at which the object was moving; this relationship is known today as the Lorentz factor. By the early 1900s, Poincaré had even begun to think that the amount of time an object experiences would change, depending on how fast the object was moving. No coherent working theory, however, was developed until 1905.

## Who finally **explained** the results of the **Michelson-Morley experiment** with a **working theory**?

The German-born physicist Albert Einstein (1879–1955) explained the Michelson-Morley experiment. In 1905—sometimes called Einstein's "year of miracles"—he published a series of scientific discoveries that forever changed the entire scientific view of the universe. He explained a biological phenomenon called Brownian motion, the electromagnetic phenomenon called the photoelectric effect, and the results of the Michelson-Morley experiment. For this he devised a new "special theory" of relativity, showing that matter and energy were related by the equation $E = mc^2$.

# TIME, WAVES, AND PARTICLES

## What is **space**?

Most people think of space as merely the absence of anything else—the "nothing" that surrounds objects in the universe. Actually, space is the fabric in which everything in the universe is embedded and through which all things travel. Imagine, for example, a gelatin dessert with pieces of fruit suspended within it. The fruit represents the objects in the universe, while the gelatin represents space. Space is not "nothing"; rather, it surrounds everything, holds everything, and contains everything in the universe.

Space has three dimensions, usually thought of as length (forward-and-backward), width (left-and-right), and height (up-and-down). It is possible to curve space, though, so that a dimension might not represent a straight line.

## What is **time**?

Time is actually a dimension, a direction that things in the universe can travel in and occupy. Just as objects in the universe can move up and down; forward and backward; or side to side, objects can also move through time. Unlike the three spatial dimensions, however, different kinds of objects in our universe move through time in only specific directions. Mathematically, it is correct to say that matter—galaxies, stars, planets, and people—only move forward in time. Meanwhile, particles made of antimatter only move backward in time; and particles of energy—such as photons, which have no mass—do not move in time.

## What is **spacetime**?

Imagine a big sheet of flexible, stretchable fabric like rubber or spandex. This sheet is like a two-dimensional surface, which can be dimpled, bent, twisted, or poked,

> ## How do space and time relate to one another?
>
> The three dimensions of space and one dimension of time are linked together as a four-dimensional fabric called spacetime. In the early twentieth century scientists such as Alexander Friedmann (1888–1925), Howard Percy Robertson (1903–1961), and Arthur Geoffrey Walker (1909–2001) presented the modern mathematical representation of how the four dimensions are linked together; this equation is called the metric of the universe.

depending on what objects are placed on it. Spacetime can be thought of as a flexible, bendable structure just like this rubber sheet, except that it is four-dimensional and its lengths and distances are related mathematically by the Friedmann-Robertson-Walker metric.

## Who first **explained** the **relationship** between **space and time**?

The famous German-American scientist Albert Einstein (1879–1955) first realized that, in order to explain the results of the Michelson-Morley experiment, travel through space and travel through time must be intimately linked. His special theory of relativity, published in 1905, showed that the faster an object moves through space, the slower it moves through time. Einstein thought there must be a very strong connection between space and time and that this connection was essential to describe the shape and structure of the universe. He did not have the mathematical expertise, however, to show how the connection might work.

Einstein consulted his friends and colleagues to figure out the best way to proceed in his research. Aided by the discoveries of the German mathematician Georg Riemann (1826–1866), the Russian-German-Swiss mathematician Herman Minkowski (1864–

1909), and the tutelage of Hungarian-Swiss mathematician Marcel Grossmann (1878–1936), Einstein learned the mathematical formulations of non-Euclidean elliptical geometry and tensors. In 1914 Einstein and Grossmann published the beginnings of a general theory of relativity and gravitation; Einstein went on to complete the formation of the theory over the next few years.

## What is Einstein's **General Theory of Relativity**?

The main ideas in the general theory of relativity are that space and time are knit

Albert Einstein. *(Library of Congress)*

25

together in a four-dimensional fabric called spacetime, and that spacetime can be bent by mass. Massive objects cause spacetime to "dimple" toward the object (think of the way that a bowling ball set on a trampoline causes the trampoline to dimple).

In the four-dimensional spacetime of the universe, if a less massive object approaches a more massive object (for example, a planet approaches a star), the less massive object will follow the lines of curved space and be drawn toward the more massive one. Thinking of the bowling ball on the trampoline, if a marble rolls past the bowling ball and into the dimpled part of the trampoline, then the marble will fall in toward the bowling ball. According to the general theory of relativity, this is how gravity works. Newton's theory of universal gravitation, according to Einstein, is almost completely correct in describing *how* gravity works, but it was not quite complete in explaining *why* it works.

## What is Einstein's **Special Theory of Relativity**?

According to the special theory of relativity, the speed of a beam of light is the same, no matter who observes it or how the observers are moving. This means that the speed of light is the fastest speed at which anything can travel in the universe.

Furthermore, if the speed of a light beam is constant, that means that other properties of motion must change. Since speed is defined as the distance traveled divided by the elapsed time, this means that the distances and times experienced by any object will change depending on how fast it is moving. The faster you move through space, the slower you move through time.

Finally, since mass can be thought of as the amount of resistance that an object has to motion, a moving object actually has more mass than when it is standing still; the faster an object is moving, the higher the mass it has. When an object reaches the speed of light, it is no longer matter, but it becomes energy. This is represented by the famous equation $E = mc^2$.

## How do **space** and **time relate** to **matter** and **energy**?

Just as general relativity is the scientific theory that explains how space and time work, quantum mechanics is the scientific theory that explains how matter and energy work. There are many key connections between relativity and mechanics. For example, there is the conversion relation between matter and energy, $E = mc^2$. Also, since matter causes gravity, it can be said that "spacetime tells matter how to move, and matter tells spacetime how to curve," as American physicist John Archibald Wheeler (1911–) phrased it.

These two major scientific theories—general relativity and quantum mechanics—do not intersect or overlap very much in terms of what aspects of the universe they describe. In fact, describing certain physical phenomena using one theory sometimes contradicts how the phenomena are described with the other. Unifying these two great theories is one of the topics at the frontier of scientific research today.

## Is it possible for one **person** to **travel more slowly through time** than another person?

It is possible for someone to travel more slowly through time relative to others. By traveling faster than someone else (say, while on a bus or airplane), time will pass by at a slightly slower rate than compared to someone standing still. The difference in these cases, however, will be incredibly small. Even if one is flying in a jet plane for twelve hours the total time difference is less than one ten-millionth of a second compared to someone who remained on the ground. Traveling at the incredible speed of 335 million miles per hour (half the speed of light) will result in the traveler experiencing an elapsed time of 10 hours and 24 minutes for every 12 hours of someone remaining stationary. But that speed is far, far beyond what our current transportation technologies can provide.

## What are **gamma rays**?

Gamma rays are electromagnetic waves whose wavelengths are shorter than about $10^{-9}$ (one ten-billionth) of a meter. These rays are very energetic and penetrative, so they can cause substantial radiation injuries to humans. Gamma rays are usually produced by the most powerful processes in the universe, such as exploding stars and supermassive black hole systems.

## What are **X rays**?

X rays are electromagnetic waves whose wavelengths range between about $10^{-9}$ and $10^{-8}$ (one ten-billionth and one hundred-millionth) of a meter. This kind of radiation can penetrate the tissues of the human body, so it may be used to take pictures of people's internal systems and skeletons, for example, at a doctor's office.

## What are **ultraviolet rays**?

Ultraviolet rays are electromagnetic waves whose wavelengths range between about $10^{-8}$ and $3.5 \times 10^{-7}$ meters. This is the kind of radiation that causes suntans and sunburns on human skin.

## What are **visible light waves**?

Visible light waves are electromagnetic waves whose wavelengths range between about $3.5 \times 10^{-7}$ and $7 \times 10^{-7}$ meters. This is the kind of electromagnetic radiation that human eyes can detect; it can roughly be divided into seven colors: violet, indigo, blue, green, yellow, orange, and red.

## What are **infrared waves**?

Infrared waves are electromagnetic waves whose wavelengths range between about $7 \times 10^{-7}$ and $10^{-4}$ meters. Humans cannot see this kind of radiation, but they can sense it as heat. Because of our warm body temperature we produce radiation mostly in the form of infrared waves. That is how some kinds of "night-vision goggles" work: they detect infrared waves coming from objects and people, even when there is not enough visible light for humans to see well.

## What are **microwaves**?

Microwaves are electromagnetic waves whose wavelengths range between about 0.0001 and 0.01 meters. This kind of radiation can be used to heat water, such as in microwave ovens, or for wireless communications, such as in cellular telephones. Microwaves are also emitted by the universe itself. The residual heat from the

> ## What is the difference between electromagnetic waves and electromagnetic radiation?
>
> **E**lectromagnetic waves and electromagnetic radiation are the same thing, but the terms are used in different contexts. Electromagnetic forces, as carried by photons, can be considered either to be waves emanating outward from a source, or as particles traveling outward from a source.

beginning of the universe leaves deep space at a temperature of about 2.7 Kelvin (2.7 degrees above absolute zero), which causes space to emit microwave radiation.

## What are **radio waves**?

Radio waves are electromagnetic waves whose wavelengths are longer than about 0.01 meters. On Earth, they can be used for communications, such as for radio or television broadcasts. In the cosmos, they are produced in large amounts by strong electromagnetic fields, fast-moving charged matter, or even by clouds of interstellar hydrogen gas.

# QUANTUM MECHANICS

## How can **light** be both a **particle** and a **wave**?

Light can be represented either as unit particles (photons) or unit waves. This phenomenon, known as wave-particle duality, is a fundamental tenet of quantum mechanics, which describes the motion of particles on very small size scales.

## What is **quantum mechanics**?

Quantum mechanics is a theory that describes the motion and behavior of matter and energy on microscopic scales. The physical laws that describe how stars, planets, and people move around in the universe simply do not work when dealing with atoms, molecules, and subatomic particles. Some of the basic concepts of quantum mechanics include:

—*Wave-particle duality:* Light is both a wave and a massless particle. Particles with mass can also be thought of as "matter waves." As a consequence, even though photons have no mass, they do have momentum and can produce force. This is very different from Newton's laws of motion, which require that objects have mass in order to have momentum and force.

—*Discrete positions and motion:* On very tiny scales, matter cannot be in every possible location. Rather, in the vicinity of any particle (for example, an atomic nucleus), other particles can only be in certain locations and at certain distances, dictated by the properties of each particle. One way to think about it is to imagine a person who is going up or down a flight of stairs, but it is only possible to stand at the heights where there are steps, and not in mid-air between two steps. Again, this is different from Newton's laws, where objects can be any distance from one another as long as the right amount of momentum or force is present.

—*Uncertainty and fluctuation:* On tiny scales, it is not possible to measure the motion or the energy of any particle at any given location or time with perfect accuracy. In fact, the more precisely a location or a time interval is measured, the less precisely known is the amount of motion or energy. That means that, for example, large flashes of energy could appear and disappear in tiny amounts of time (much, much less than a trillionth of a second!), and we would never notice because the time interval is too short for us to observe the flashes. Scientists hypothesize that a drastic energy fluctuation of this kind might have occurred at the birth of our universe—the Big Bang.

### Who was **Max Planck** and what did he contribute to our understanding of matter and **quantum mechanics**?

German physicist Max Planck (1858–1947) figured prominently in the development of modern physics, especially in the field of quantum mechanics. As he studied how thermal radiation—elec-

Max Planck. *(Library of Congress)*

tromagnetic waves emitted by hot objects—worked, Planck was the first to derive a mathematically correct way to describe the spectral distribution of energy from a thermally emitting object. To do that, however, Planck used a mathematical method that suggested that light was comprised not of continuous waves, but of particles or "pieces" of light called quanta. His theory soon proved to be a fundamental property of light. Today, the main research organization in his native Germany is known as the Max Planck Society, and the national laboratories of natural sciences in Germany are called the Max Planck Institutes in his honor.

Ernest Rutherford. *(Library of Congress)*

### Who was **Ernest Rutherford** and what did he contribute to our understanding of matter and **quantum mechanics**?

New Zealand physicist Ernest Rutherford (1871–1937) contributed greatly to the understanding of matter, especially its microsocopic structure, and to the understanding of radioactivity. Rutherford is credited with creating the names "alpha," "beta," and "gamma" rays to describe different kinds of radioactive emissions. In his most famous experiment, he tried to figure out the structure of atoms by firing radioactive particles at a thin sheet of gold atoms. He expected the particles to be slightly deflected by the atoms; instead, to his surprise, very few particles were deflected at all, and some of them bounced right back as if they had hit a solid wall. Rutherford interpreted this result to mean that atoms consist of a large volume of emptiness, occupied by tiny negative charges, and a very small but very dense nucleus that contained positive charge. Rutherford's experimental result was the strongest evidence that matter is actually built of atoms.

### How did **Albert Einstein** contribute to our understanding of matter and **quantum mechanics**?

In 1905 Albert Einstein not only published his special theory of relativity, but also two other theories that became part of the fundamental understanding of matter in the universe. In one of the two theories, he explained that Brownian motion—the seemingly random jiggling motions of microscopic fat globules suspended in milk or water—were caused by individual atoms and molecules moving around the suspension, striking the globules and causing them to move. In the other theory, he explained that the photoelectric effect—in which light of certain colors striking sheets of metal would produce electric currents, whereas light of other colors would not—was caused by light acting as both a wave and a particle. The Brownian motion

## Who eventually helped settle the theory of quantum mechanics, and when did it happen?

**M**ost scientists agree that it was not until about 1937 that quantum mechanics was finally considered the correct way to describe the behavior of matter and energy on microscopic scales. Scientists such as the English physicist Paul Dirac (1902–1984), German physicist Wolfgang Pauli (1900–1958), French physicist Louis de Broglie (1892–1987), Austrian physicist Ernest Schroedinger (1887–1961), and the German physicist Werner Heisenberg (1901–1976) all worked to help establish the mathematical framework for the theory and decipher the details of quantum phenomena. Overall, no single person can be credited with the discovery of quantum mechanics. As with many triumphs of science, many brilliant people worked for a very long time to figure out how it all came together.

result further helped prove the existence of atoms; and the photoelectric effect result showed that new physical ideas, such as quantum mechanics, were necessary to explain the nature and behavior of light.

## How has **quantum mechanics** advanced **in recent years**?

As with all important scientific theories, quantum mechanics has advanced a great deal since its initial formulation and confirmation. The original quantum theory has advanced to the point today where scientists have described what the standard models of subatomic particles in the universe are (such as fermions, bosons, quarks, leptons, and so forth) and their very complex behaviors and interactions (quantum electrodynamics, quantum chromodynamics, and more). The fundamental nature of matter and energy is still being studied today, and many more exciting discoveries and advances are sure to come in the future.

# CHARACTERISTICS OF THE UNIVERSE

### What is **the universe**?

The universe is all of space, time, matter, and energy that exist. Most people think of the universe as just space, but space is just the framework, the "scaffolding" in which the universe exists. Furthermore, space and time are intimately connected in a four-dimensional fabric called spacetime.

Amazingly, some hypotheses suggest that the universe we live in is not all there is. In this case, there is more than just space, time, matter, and energy. Other dimensions exist, and possibly other universes. None of those models, however, have yet been confirmed.

### Why **does** the **universe exist**?

That, for better or worse, is not answerable by science alone. Astronomy can describe, however, a theory that explains how the universe began.

### How **old** is the **universe**?

The universe is not infinitely old. According to modern astronomical measurements, the universe began to exist about 13.7 billion years ago.

### Is the **universe infinite**?

It has not yet been scientifically determined exactly how large the universe is. It may indeed be infinitely large, but we have no way yet to confirm this possibility scientifically.

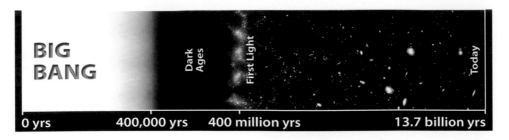

BIG BANG

Dark Ages

First Light

Today

| 0 yrs | 400,000 yrs | 400 million yrs | 13.7 billion yrs |

Scientists estimate that the universe is about 13.7 billion years old. (*NASA/JPL-Caltech/A. Kashlinsky*)

## What is the **structure** of the **universe**?

The structure of the universe—as opposed to the structure of matter in the universe—is determined by the shape of space. The shape of space is, surprisingly, curved. On a very large scale—millions or even billions of light-years across—space has a three-dimensional "saddle shape" that mathematicians refer to as "negative curvature." In our daily lives, however, it is such a tiny effect that we do not notice it.

On smaller scales—that of planets, stars, and galaxies—the structure of the universe can be altered by massive objects. This alteration manifests itself as the curvature of space and time, as explained by the general theory of relativity.

## How **big** is the **universe**?

Here on Earth, in the Milky Way galaxy, there is a limit to how far out into the universe humans can observe, regardless of what technology is used. Imagine, for example, being on a ship in the middle of the ocean. If you look in all directions, all you see is water, out to a certain distance. But Earth's surface extends far beyond that horizon limit. The farthest limit to our viewing is called the cosmic horizon, which is about 13.7 billion light-years away, or about 80 billion trillion miles, in every direction. Everything within that cosmic horizon is called the observable universe. In many cases, for the sake of brevity, astronomers refer to the "observable universe" as merely the "universe."

As for the universe beyond the cosmic horizon, there is still no scientific way to measure its size. There is no reason to think there is or is not a boundary far away. However, it is possible for the universe to be limited in size and still not have an edge. Think of the surface of our planet, for example. Earth's surface area is finite, but there is nowhere on Earth where you could reach the "end" of Earth in a boat and fall off our planet. In a huge, three-dimensional way, our universe, might be similar.

## What are the **possible shapes** of the **universe**?

There are three general categories of possible shapes of the universe: open, flat, and closed. These adjectives refer to the kind of curvature that space has overall. Massive objects cause space to bend and curve; the universe itself is a massive object, so the entire cosmos is curved, too.

## What is the difference between a **closed universe**, a **flat universe**, and an **open universe**?

Closed universe—A closed universe curves in upon itself, so that its total volume could be limited. A two-dimensional example might be the surface of a sphere; there is no sharp edge, but the overall shape is bounded. As a closed universe expands, the edges of any given volume of space "pinch" inward, so that the expansion will ultimately end and might reverse into a contraction called the Big Crunch.

The universe is incredibly vast, stretching out billions of light-years in every direction and containing billions of galaxies. (*NASA/JPL-Caltech/A. Kashlinsky*)

Flat universe—A flat universe has no net curvature. A two-dimensional example might be the surface area of a cube. All the small curves caused by massive objects average out to zero; length, width, and height are straight lines and extend all the way across the universe. As a flat universe expands, the edges of any given volume of space will stay straight, and the expansion will continue indefinitely.

Open universe—An open universe curves outward, so that its total volume cannot be limited. A two-dimensional example might be an equestrian saddle; the curvature bends away from the center of the shape, and would keep going without limit if the surface were extended. As an open universe expands, the edges of any given volume of space "bow" outward, so the expansion will not end.

# ORIGIN OF THE UNIVERSE

## How did the **universe begin**?

The scientific theory that describes the origin of the universe is called the Big Bang. According to the Big Bang theory, the universe began to exist as a single point of spacetime, and it has been expanding ever since. As that expansion has occurred, the conditions in the universe have changed—from small to big, from hot to cold, and from young to old—resulting in the universe we observe today.

## Who were the **first scientists** to **formulate** the **Big Bang theory**?

In 1917 Dutch astronomer Willem de Sitter (1872–1934) showed how Albert Einstein's general theory of relativity could be used to describe an expanding universe. In 1922, Russian mathematician Alexander Friedmann (1888–1925) derived an exact mathematical description of an expanding universe. In the late 1920s, the Belgian astronomer Georges-Henri Lemaître (1894–1966) independently rediscovered Friedmann's mathematical formulation. Lemaître deduced that if the universe were indeed expanding, and has been doing so for its entire existence, then there would have to be a moment in the distant past when the whole universe occupied just a single point. That moment, and that point, would be the origin of the cosmos. Lemaître's work, and that of de Sitter and Friedmann, were eventually confirmed through observations; since Lemaître was a Jesuit priest as well as an astronomer, he has sometimes been called "the father of the Big Bang."

## Who developed the idea of a **"hot"** Big Bang?

The Russian-born American physicist George Gamow (1904–1968) furthered the Big Bang model by including the distribution of energy in the universe. If such a bang had occurred, he argued, the universe would have been incredibly hot very soon after the bang—somewhere in the area of trillions upon trillions of degrees. As the universe expanded, the heat in the universe would become distributed over a larger volume, and the temperature would go down. After one second, the average cosmic temperature would drop to about a billion degrees; after half a million years, the average temperature would be a few thousand degrees; and so on. Even after billions of years had passed, however, Gamow showed that this background heat would persist. After about 15 billion years, it would appear as a background radiation field that would be just a few degrees above absolute zero. Gamow predicted that this cosmic background radiation could be detected by its microwave radiation. In 1965 the cosmic microwave background radiation was indeed discovered.

## Is the **Big Bang** a **theory** or a **fact**?

It is a theory. Scientifically speaking, that makes it more powerful than fact. Facts are single pieces of information, while theories incorporate many, many facts into a conceptual model, which is then confirmed by a process of prediction, observation, and experiment. In science, individual facts can be weak and often turn out to be wrong, whereas theories are not easily disproved and are strongly supported by evidence.

The Big Bang theory has solid scientific evidence to support it, and its fundamental concepts have been scientifically proven to be correct. However, like all major theories in science, there are many details yet unproven and many questions still unanswered. These many important unknowns will continue to lead scientists to search for answers and make new discoveries, as they try to understand the cosmos.

## According to the **Big Bang theory**, what happened when the **universe began**?

The Big Bang theory does not explain why the Big Bang actually happened. A well-established hypothesis is that the universe began in a "quantum foam"—a formless void where bubbles of matter, far smaller than atoms, were fluctuating in and out of existence on timescales far shorter than a trillionth of a trillionth of a trillionth of a second. In our universe today, such quantum fluctuations are thought to occur,

## What was there before the Big Bang?

It is not scientifically possible to ask what came "before" the Big Bang. That is because time itself did not exist until the Big Bang occurred. Just as there is nothing "north" of the North Pole, because Earth does not extend any farther north, there was nothing "before" the first instant of time.

However, if you can imagine the existence of more than one universe—a possible consequence of membrane theory or string theory—then it is possible that other universes, with other dimensions of space and time, could have existed before the universe.

but they happen so quickly that they never affect what happens in the cosmos. But if, 13.7 billion years ago, one particular fluctuation appeared but did not disappear, suddenly ballooning outward into a gigantic, explosive expansion, then it is possible that something like today's universe could have been the eventual result.

In another, more recently proposed hypothesis, the universe is a four-dimensional spacetime that exists at the intersection of two five-dimensional structures called membranes. Picture two soap bubbles that come in contact with one another and stick together: the "skin" where the bubbles intersect is a two-dimensional result of the interaction between two three-dimensional structures. If the membrane hypothesis is correct, then the Big Bang event marked the moment the two membranes made contact. Neither of these models has any kind of experimental or observational confirmation yet.

### How close to the **initial instant** of the Big Bang can the **behavior of the universe** be **traced**?

The Big Bang event itself is a singularity, where (and when) the currently understood laws of physics cannot describe what is going on. This means that the behavior of the universe can only be traced back to a time after the Big Bang, when the laws of physics first begin to apply. By combining the minimum size and time scales that are described by the two major theories that describe the universe—general relativity and quantum mechanics—scientists have deduced that the earliest time that the behavior of the universe can be traced is about $10^{-43}$ seconds after the Big Bang. That is a ten-millionth of a trillionth of a trillionth of a trillionth of a second! This earliest, all-but-unknowable period of cosmic history is called the Planck time, after the German physicist, and pioneer of quantum theory, Max Planck.

### What was the **size** of the **universe** at the **Planck time**?

The size of the universe at the Planck time was approximately the distance light can travel within that time interval. That means that the diameter of the universe was

$10^{-35}$ meters, or about a billionth of a trillionth of a trillionth of an inch. This length is known as the Planck length.

### How **massive** and how **dense** was the **universe** at the **Planck time**?

By using much of the same reasoning with which the Planck time and Planck length are derived, it is also possible to calculate the mass and density of the universe at the Planck time. It turns out that the Planck mass minus the mass of the universe $10^{-43}$ seconds after the Big Bang is about a thousandth of a milligram.

That does not sound like much by terrestrial standards. Remember, though, that mass is contained within a volume that is less than a hundredth of a billionth of a billionth the diameter of an atomic nucleus. So the density of that primordial universe is an incredible $10^{94}$ times the density of water. Nothing in our universe today that we know of, including the densest known black holes, even remotely approaches that kind of density. Energy this concentrated must certainly behave in ways almost unimaginable in the current universe, and that is almost certainly reflected in everything that happened in the infant cosmos.

### When and how did **matter first form**?

After the Planck time, the universe expanded rapidly, and all the energy rushed outward to fill that expanding volume; as a result, the universe began to cool down. By about a millionth of a second after the Big Bang, the temperature of the universe was still well above a trillion degrees. But the energy density had probably dropped enough that subatomic particles of matter could come into existence for brief periods of time, reverting back and forth between matter and energy. This state of the universe, informally called the "quark-gluon soup," may not even be the earliest appearance of matter in the universe. Still, it is the hottest and earliest cosmic state scientists have been able to simulate so far, using huge supercolliders that can generate microscopic bursts of tremendous energy density.

## Why is the **inflationary model important** in the modern Big Bang theory?

The inflationary model was proposed in the early 1970s to explain two key observations about the universe. First, the matter and energy in the universe appears to be statistically the same in every direction, as far as astronomers are able to observe. This means that parts of the universe that do not share a cosmic horizon today—that is, parts that should not have to be the same— somehow shared a cosmic horizon long ago in the past. (This is called the "horizon problem.") Second, the geometry of the universe is remarkably close to "flat" when, again, there is no reason why such a special geometry should exist. (This is called the "flatness problem.")

An artist's conception of the universe rapidly expanding after the Big Bang. (*iStock*)

As it is currently modeled, the inflationary period in the early universe addresses both the horizon problem and the flatness problem. The hyperinflation was so fast that it carried parts of space that used to share a cosmic horizon away from each other, so that in the present universe they would be statistically identical, even though they were no longer close enough to achieve balance with one another. In addition, the hyperinflation happened in such a way that it forced all of space to become a "flat" geometric structure. Although the model seems to explain what has been observed about the universe, it does not explain why it happened, nor exactly how much larger the universe grew during that time.

# EVIDENCE OF THE BIG BANG

## What **evidence** is there for the **Big Bang** based on the **motion of objects** in the universe?

The expanding universe is a solid piece of observational evidence that the universe began as the Big Bang theory describes. If space is getting larger all the time, then that means the universe is larger now than it was yesterday. Similarly, it was larger yesterday than it was last month, and larger last month than last year. By continuing backward in time, it is possible to follow the trend all the way down to the point where the entire universe was just a point. Based on the expansion rate of the universe, that point was at a time about 13.7 billion years ago.

## Who discovered the cosmic microwave background?

In the 1960s, astronomers Arno Penzias (1933–) and Robert Wilson (1936–) were conducting research at Bell Telephone Laboratories in Holmdel, New Jersey. For their telescope they were using a very sensitive, horn-shaped antenna that was originally developed to receive weak microwave signals for use in wireless communications. While they were testing this antenna, Penzias and Wilson detected a ubiquitous microwave static that came from all directions in the sky. After examining four years of data, and checking carefully to be sure there was no interference or malfunction in the equipment, they interpreted their "static" as a real signal, coming from outer space in every direction. After consulting with astrophysics colleagues at Princeton, they realized that they had indeed detected the cosmic microwave background. They published their results in 1965, and it was immediately recognized as scientific evidence confirming the Big Bang theory.

### What **evidence** is there for the **Big Bang** based on the **nature of matter** in the universe?

The distribution of elements by mass in the early universe—75 percent hydrogen, 25 percent helium, and a tiny trace of other, heavier elements—matches the predictions of a hot Big Bang. This kind of elemental distribution probably came about because, as the universe cooled and expanded, there was only a very short amount of time (about three minutes!) when conditions in the universe could support the creation of atomic nuclei from subatomic particles.

### What **evidence** is there for the **Big Bang** based on the **nature of energy** in the universe?

Perhaps the most convincing evidence confirming the Big Bang theory is the cosmic microwave background radiation: the leftover energy from the hot, early universe that still fills space and permeates the cosmos in every direction. Scientists had predicted that such background radiation would indicate that the temperature of space would be several degrees above absolute zero. The detection of the background radiation showed that the temperature was very close to 3 degrees Kelvin— a spectacular success of the scientific method.

### What **follow-up study** of the **cosmic microwave background** solidly **confirmed** the **Big Bang** theory?

In 1992 NASA launched the Cosmic Background Explorer (COBE) satellite. Its purpose was to study the nature of the cosmic microwave background radiation. Instruments on COBE confirmed that the radiation detected by Penzias and Wilson

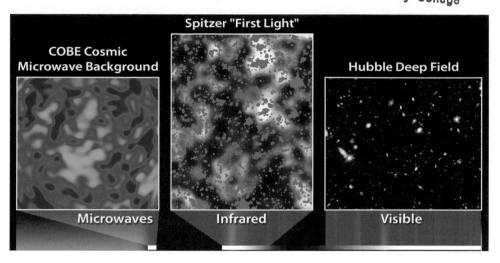

Three views of the universe from observatories show the stars and galaxies as they appear in the visible light spectrum, in infrared, and how the microwave background appears. This microwave background provides evidence of the Big Bang, according to astronomers. (*NASA/JPL-Caltech/A. Kashlinsky*)

in 1967 was a nearly perfect profile of the temperature of the universe, and that the cosmic microwave background temperature was almost exactly 2.73 degrees Kelvin (about 454.7 degrees below zero Fahrenheit). Furthermore, careful analysis showed that there are tiny variations of temperature in the background radiation: these variations were barely a few ten-thousandths of a degree Kelvin and are the fossilized signature of the original miniscule fluctuations of the matter and energy density in the early universe from nearly 13.7 billion years ago. Those fluctuations seeded the changes that have since caused the universe to age and evolve from what it once was—a kernel of spacetime nearly uniformly filled with energy—into what it is today—a vastly variegated tapestry of dense and sparse regions, sprinkled with galaxies, stars, planets, and more.

# EVOLUTION OF THE UNIVERSE

### Who **first showed** that the **universe is expanding**?

The same astronomer who showed that galaxies exist outside the Milky Way also demonstrated that the universe is expanding. Edwin Hubble continued to study galaxies after his pioneering measurement of the distance to the Andromeda Galaxy. He examined the relationship between the motion of a galaxy and the distance the galaxy is away from Earth. He discovered that the farther away a galaxy is, the faster it moves away from us, which is the telltale sign of an expanding universe.

### What is the **Hubble Constant**?

The expansion rate of the universe is called the Hubble Constant in honor of Edwin Hubble (1889–1953). Currently the best measured value of the Hubble Constant is about 73 kilometers per second per megaparsec. That means that, if a location in space is one million parsecs from another location, then in the absence of any other forces or effects the two locations will be moving apart from one another at the speed of 163,000 miles (263,000 kilometers) per hour.

### How did Hubble use the **Doppler effect** to **measure** the **universe**?

Hubble measured the galaxies' Doppler effect—the shift in the observed color of objects moving toward or away from an observer—by mounting a machine called a spectrograph on a telescope. He split the light from distant galaxies into its component parts and measured how far the wavelengths of emitted light shifted toward longer wavelengths.

### How does the **Doppler effect** work for **light**?

When an object emitting light—or any kind of electromagnetic radiation, for that matter—moves toward someone, the wavelength of its emitted light is decreased. Conversely, when the object moves away, the wavelength of its emitted light is increased. For visible light, the bluer part of the spectrum has shorter wavelengths, and the redder part of the spectrum has longer wavelengths. Thus, the Doppler effect for light is called a "blueshift" if the light source is coming toward an observer, and a "redshift" if it is moving away. The faster the object moves, the greater the blueshift or redshift.

### Who first **discovered** the **Doppler effect for light** from an astronomical source?

The first astronomer to observe a Doppler shift from a distant object was Vesto Melvin Slipher in 1912. Slipher (1875–1969) used telescopes to photograph and study large fuzzy patches of gas and dust, called nebulae, which were thought to be within the Milky Way galaxy. Much to everyone's surprise, Slipher found that many of these patches were made of stars, which suggested that they could be distant galaxies like the Milky Way.

## How does Hubble's original expansion rate compare to the modern value of the Hubble Constant?

It is way off—about seven times greater than the modern value. Even so, Edwin Hubble's measurement methods made sense, and his general conclusion—the correct formula that distance to an object is directly proportional to its velocity away from the observer—is a relation known today as the Hubble Law. As a result, astronomers today still give him credit for the discovery of the expanding universe.

In 1903 Slipher accepted a scientific position at the Lowell Observatory in Flagstaff, Arizona. He was brought to Flagstaff by the astronomer Percival Lowell to investigate these nebulae. Lowell thought that some of these cloud-like structures, particularly the ones that had spiral patterns, might be the beginnings of other solar systems within our galaxy. Slipher's job was to study the spectra of the nebulae so they could be carefully analyzed.

Studying the remarkable spectrum of the Andromeda nebula, Slipher discovered it did not match the spectrum of any known gas. Rather, it was more like the spectrum made by starlight. Even more amazing, the colors of that starlight appeared to be blueshifted. Slipher concluded that the Andromeda nebula was actually moving toward Earth at a remarkable speed of about half a million miles per hour. Over the following years, Slipher analyzed the spectra of 12 other spiral nebulae. He found that some were moving toward Earth and some were moving away. Furthermore, these nebulae were moving at remarkable speeds of up to 2.5 million miles per hour (1,100 kilometers per second). He concluded that these objects were not nebulae at all, but entire systems of millions or billions of stars, so distant that they had to be galaxies. Slipher's pioneering work was later confirmed by Edwin Hubble, who used Cepheid variables as standard candles to prove that the great nebula in Andromeda was in fact the Andromeda Galaxy.

## What **rate of expansion** did Edwin **Hubble measure** for the universe?

Edwin Hubble's original measurement of the expansion rate of the universe was about 500 kilometers per second per kiloparsec.

## Aside from cosmic expansion, what **other forces** might **move objects** in the universe?

Aside from being carried along by cosmic expansion, the only other force in the universe that can make large objects like planets, stars, and galaxies move large distances is gravity.

### What is the **universe expanding into**?

The whole universe is expanding. That means that all of space is expanding, and every location in space is moving away from every other location in space, unless there is mass nearby creating gravity. So our three-dimensional space cannot be expanding into another three-dimensional space.

One way to think of this is to imagine a balloon being inflated. The balloon is a curved, two-dimensional piece of elastic rubber that is expanding as it is inflated. It is not expanding into another two-dimensional surface, however. Rather, the balloon is expanding into a three-dimensional space. By adding one dimension to this example, the result is a three-dimensional space that can be thought of as expanding into an extra dimension. In the case of the universe, this is the four-dimensional spacetime that comprises the cosmos.

# BLACK HOLES

### What **objects** in the universe have the **strongest gravity**?

The most massive objects in the universe exert the most gravity. However, the strength of a gravitational field near any given object also depends on the size of the object. The smaller the object, the stronger the field. The ultimate combination of large mass and small size is the black hole.

### What is a **black hole**?

One definition of a black hole is an object whose escape velocity equals or exceeds the speed of light. The idea was first proposed in the 1700s, when scientists hypothesized that Newton's law of universal gravitation allowed for the possibility of stars that were so small and massive that particles of light could not escape. Thus, the star would be black.

### What does **relativity** have to do with **black holes**?

The idea of black or dark stars was interesting, but it was not explored scientifically for more than a century after the notion was proposed in the 1700s. After 1919, when the general theory of relativity was confirmed, scientists started to explore the implications of gravity as the curvature of space by matter. Physicists realized that there could be locations in the universe where space was so severely curved that it would actually be "ripped" or "pinched off." Anything that fell into that location would not be able to leave. This general relativistic idea of an inescapable spot in space—a hole where not even light could leave—led physicists to coin the term "black hole."

### How can **astronomers find black holes** if they cannot see them?

The key to finding black holes is their immense gravitational power. One way to find black holes is to observe matter moving or orbiting at much higher speeds than

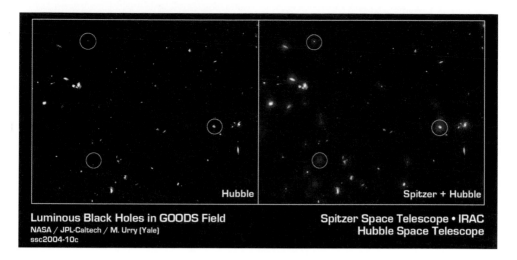

Luminous Black Holes in GOODS Field
NASA / JPL-Caltech / M. Urry (Yale)
ssc2004-10c

Hubble

Spitzer + Hubble

Spitzer Space Telescope • IRAC
Hubble Space Telescope

Astronomers can detect black holes by searching for sources of X-ray radiation. The image compiled using data from the Spitzer Space Telescope and Hubble Space Telescope shows X-ray sources indicating black holes; the image on the right shows the same section of space in normal, visible light. (*NASA/JPL-Caltech/Yale*)

expected. By carefully mapping the motion, and then applying Kepler's third law of orbital motion, it is possible to measure the mass of an object even without seeing the object itself.

The deep gravitational field of a black hole can also produce a tremendous amount of light nearby and around itself, even if the hole itself is dark. Matter falling into a black hole runs into a lot of other material that has collected around the hole. Just as a meteorite or spacecraft gets hot as it enters Earth's atmosphere, the infalling matter gets hot from the frictional drag too, sometimes reaching temperatures of millions of degrees. That hot material glows brightly and emits far more X-ray radiation and radio waves than would normally be expected from such a small volume of space. By searching for these tell-tale emissions, astronomers can deduce the presence of black holes even though they cannot see the holes themselves.

## What **kinds of black holes** are there?

Two categories of black holes are known to exist, and a third kind has been hypothesized but not yet detected. One kind, known loosely as a stellar black hole or low-mass black hole, is found wherever the core of a very massive star (usually 20 or more times the mass of the Sun) has collapsed. The other known kind, called a supermassive black hole, is found at the centers of galaxies and is millions or even billions of times more massive than the Sun.

The third kind of black hole, called a primordial black hole, is found at random locations in space. It is hypothesized that these black holes were created at the beginning of cosmic expansion as little "imperfections" in the fabric of spacetime. However, no such black hole has yet been confirmed to exist.

## What is the **structure** of a **black hole**?

The center of the black hole—the actual "rip" or "pinch" in the fabric of space-time—is called the singularity. It is a single point that has no volume but infinite density. Amazingly, the laws of physics as we understand them simply do not work at the singularity of a black hole the way they do in the rest of the universe.

Surrounding the singularity is a boundary called the event horizon. This is the place of no return, where the escape velocity for the black hole is the speed of light. The more massive the black hole is, the farther the event horizon is from the singularity, and the larger the black hole is in size.

## Can **anything escape** a **black hole**?

According to British physicist Stephen Hawking (1942–), energy can slowly leak out of a black hole. This leakage, called Hawking radiation, occurs because the event horizon (boundary) of a black hole is not a perfectly smooth surface, but "shimmers" at a subatomic level due to quantum mechanical effects. At these quantum mechanical scales, space can be thought of as being filled with so-called virtual particles, which cannot be detected themselves but can be observed by their effects on other objects. Virtual particles come in two "halves," and if a virtual particle is produced just inside the event horizon, there is a tiny chance that one "half" might fall deeper into the black hole, while the other "half" would tunnel through the shimmering event horizon and leak back into the universe.

## What does **Hawking radiation do** to black holes?

Hawking radiation is a very, very slow process. A black hole with the mass of the Sun, for example, would take many trillions of years—far longer than the current age of the universe—before its Hawking radiation had any significant impact on its size or mass. Given enough time, though, the energy that leaks through a black hole's event horizon becomes appreciable. Since matter and energy can directly convert from one to another, the black hole's mass will decrease a corresponding amount.

According to theoretical calculations, a black hole having the mass of Mount Everest—which, by the way, would have an event horizon smaller than an atomic nucleus—would take about 10 to 20 billion years to lose all its energy, and thus matter, back into the universe due to Hawking radiation. In the final instant, when the last bit of matter is lost, the black hole will vanish in a violent explosion that may release a huge blast of high-energy gamma rays. Perhaps astronomers may someday observe just such a phenomenon and confirm the idea of Hawking radiation as a scientific theory.

Physicist Stephen Hawking (seated) visits the CERN particle physics laboratory in Geneva, Switzerland. Hawking first theorized that black holes could emit radiation that would, after billions of years, eventually cause the black hole to disappear. (*M. Brice, CERN*)

## What happens when a **black hole spins**?

When a black hole rotates, the shape and structure of the event horizon can change. If a black hole is not rotating, then the event horizon is a perfect sphere centered on its singularity. As a black hole spins, the event horizon flattens into a sort of thick doughnut shape, and a structure called the ergosphere can develop. Here, light beams do not escape the black hole but instead orbit around the singularity.

## What happens if a **spinning black hole** has **electric charge**?

When charged particles spin around and around, electromagnetic fields are produced. Since black holes contain large amounts of mass in small volumes, the speed of the spin can be enormous, and the density of the electrical charge can be enormous as well. The combination produces some of the strongest magnetic fields anywhere in the universe.

In this situation, when matter falls toward the black hole it not only gets superhot but also supermagnetized. While much of the falling material disappears into the black hole never to be seen again, some of it will be channeled into the magnetic fields and fired outward along superpowerful, magnetically focused jets. Depending on how massive and strongly charged the black hole is, these jets could propel matter into space at 99 percent the speed of light or more and extend for thousands or even millions of light-years. These relativistic jets emanating from black hole systems are some of the most dramatic structures in the cosmos.

## How **big** are **black holes**?

The singularity at the center of any black hole has no volume. The size of the event horizon—the boundary of no return—of a black hole, on the other hand, varies depending on the black hole's mass. The mathematical relationship between the mass of a black hole and the size of its event horizon was derived by the German astrophysicist Karl Schwarzschild (1873–1916). The radius of a black hole's event horizon is named the Schwarzschild radius in his honor.

Generally speaking, the Schwarzschild radius of a stellar black hole is about a hundred miles, while the Schwarzschild radius of a supermassive black hole ranges from a few million to a few billion miles. (For reference, the average distance between the Sun and Pluto is about three billion miles.) If the Sun were squeezed small enough to become a black hole, its Schwarzschild radius would be about three miles; and if Earth were squeezed small enough to become a black hole, its Schwarzschild radius would be about three-quarters of an inch.

## Where are some **black holes located** in our galaxy?

The table below lists some known black holes in the Milky Way galaxy.

**Black Holes in the Milky Way Galaxy**

| Name | Probable Mass in solar masses | Approx. Distance from Earth in light-years |
|---|---|---|
| A0620-00 | 9–13 | 3,000–4,000 |
| GRO J1655-40 | 6–6.5 | 5,000–10,000 |
| XTE J1118+480 | 6.4–7.2 | 6,000–6,500 |
| Cygnus X-1 | 7–13 | 6,000–8,000 |
| GRO J0422+32 | 3–5 | 8,000–9,000 |
| GS 2000-25 | 7–8 | 8,500–9,000 |
| V404 Cygnus | 10–14 | 10,000 |
| GX 339-4 | 5–6 | 15,000 |
| GRS 1124-683 | 6.5–8.2 | 17,000 |
| XTE J1550-564 | 10–11 | 17,000 |
| XTE J1819-254 | 10–18 | less than 25,000 |
| 4U 1543-475 | 8–10 | 24,000 |
| Sagittarius A* | 3,000,000 | Center of the Milky Way |

## What would happen to a person who fell into a black hole?

That depends on the size of that black hole. If a human fell into a small black hole with a high density, very strong tidal forces would cause substantial physical destruction to his or her body. The front of the body would be accelerated so much more vigorously than the rear that the atoms and molecules would be pulled apart from one another. The unfortunate person would flow into the black hole as a stream of subatomic particles.

However, if someone fell into a supermassive black hole with a low density, there would be no such tidal forces to contend with. In that case, the relativistic effects of being near the event horizon of a black hole would become apparent. As the person fell closer and closer to the event horizon, his or her speed would get closer and closer to the speed of light, and the faster the speed, the slower one moves through time. Eventually, the person would move so slowly through time that he or she would effectively freeze, never reaching the event horizon. In fact, the event horizon will grow outward to meet the person. As it did so, the person's body would change from matter into energy, following the formula $E = mc^2$, and disappear into the black hole forever.

## How **dense** are **black holes**?

Based on Karl Schwarzschild's formula for the radius of a black hole's event horizon, the density of a black hole depends very strongly on its mass. A black hole the mass of Earth, for instance, would have more than 200 trillion trillion times the density of lead. On the other hand, a black hole that is one billion times the mass of the Sun would have an average density much lower than the density of water.

## Could a **giant black hole** someday **consume** the entire **universe**?

No, a giant black hole will not consume our universe. Remember, black holes are deep gravitational structures, not cosmic "vacuum cleaners," and they do not "suck" things in. Picture a manhole in a construction zone on a busy city sidewalk: if someone falls in, he or she may not come back out again, but if one avoids the hole and its environment then there is no danger. That is the same way a black hole works. No matter how massive a black hole is, there is always a limit to its gravitational influence, and matter that lies beyond that limit will not be affected by the black hole at all.

# WORMHOLES AND COSMIC STRINGS

## What is a **wormhole**?

According to current hypotheses, a wormhole is an imperfection in spacetime that has two ends. Instead of a black hole, which only has one point singularity in space-

Mathematically, it is possible to manipulate the equations of Einstein's general theory of relativity to create a wormhole that could stretch across a large distance in space. Then, if the known laws of physics do not apply in the wormhole, it might be mathematically possible to go from one end to the other in an amount of time shorter than a beam of light would take to traverse that same distance. However, those same manipulated equations suggest that nothing larger than microscopic particles could get through a wormhole without being destroyed by the extreme conditions within.

time, a wormhole could have one point where matter can only enter and another point where matter can only exit.

### Do **wormholes** really **exist**?

No wormhole has ever been detected. Science fiction writers like to invoke wormholes as useful ways to violate the known laws of physics (for instance, making objects disappear into nothingness or appear out of nowhere, for no apparent reason), but real wormholes, if they exist, would destroy anything on a terrestrial scale that is near one of its openings.

### What is a **cosmic string**?

According to current hypotheses, a cosmic string is a giant vibrating strand or closed loop of matter; it is almost like a black hole, but long and thin, rather than a point or sphere. Cosmic strings may have been produced by gravitational shifts in the early universe. They could be envisioned as "creases" left in an otherwise smooth transition from the initial phases of cosmic evolution. They might also be described as "wrinkles" in the texture of the universe, moving and wiggling around in spacetime. A cosmic string may be many light-years long, but far thinner than the width of a human hair, and may contain the mass of billions upon billions of stars. A cosmic string may also carry an extremely strong electrical current.

### Do **cosmic strings** really **exist**?

No cosmic string has ever been detected. Every once in a while, observational evidence suggests that a cosmic string might have been seen, but these observations have never been confirmed. It may be possible that the universe may have contained many cosmic strings early in its history, but almost all of them may have decayed away by now.

### Can **cosmic strings** be used to **travel backward** in **time**?

The American astrophysicist J. Richard Gott (1938–) has published a book describing a special kind of time machine that might be possible using cosmic strings. In a nutshell, if there are two straight cosmic strings passing close by to one another as they move about in the universe, the spacetime between the two strings will be heavily distorted by the strings' gravitational influence, and time could loop around in a strange configuration. If an object can somehow follow that loop in exactly the right way, it could wind up taking a wild, corkscrew path through time so that it would end up in a location in spacetime *before* where it started. Research into the theoretical possibilities of such a "Gott time machine" continues, but again, no cosmic strings have ever been detected, let alone two.

# DARK MATTER
# AND DARK ENERGY

## What is **dark matter**?

In the 1930s, astronomer Fritz Zwicky (1898–1974) noticed that, in the Coma cluster of galaxies, many of the individual galaxies were moving around so fast that there had to be a tremendous amount of gravitational pull toward the center of the cluster; otherwise, the galaxies would literally fling themselves out of the cluster. The amount of matter that needed to exist in the cluster to produce that much gravity far exceeded the amount of matter observed in all the galaxies in the cluster put together. This extra matter became known as "dark matter."

In 1970 astronomer Vera C. Rubin (1928–) and physicist W. Kent Ford showed that stars in the Andromeda Galaxy were moving so fast that for the stars to stay in the galaxy there had to be a tremendous amount of matter surrounding and enveloping the entire galaxy like a giant cocoon. Since this matter is not visible to telescopes by the light it emits, but rather only by the gravity it exerts, this, too, is an example of evidence for dark matter.

After decades of further study, dark matter has now been confirmed as an important constituent of matter around galaxies, in clusters of galaxies, and throughout the universe as a whole. According to the latest measurements, about 80 percent of the matter in the universe is dark matter.

An artist's drawing of the Spitzer observatory observing the object OGLE-2005-SMC-001, a dark body that can only be detected by analyzing light sources around it. Such objects are evidence of dark matter in the universe. (*NASA/JPL-Caltech/R. Hurt*)

51

THE UNIVERSE

## What is **dark energy**?

When Albert Einstein, Willem de Sitter, Alexander Friedmann, Georges-Henri Lemaître and others were working on the nature of the universe in the early twentieth century, Einstein introduced a mathematical term into his equations to keep a balance between cosmic expansion and gravitational attraction. This term became known as the "cosmological constant," and seemed to represent an unseen energy that emanated from space itself.

After Edwin Hubble and other astronomers showed that the universe was indeed expanding, the cosmological constant no longer appeared to be necessary, and so it was not seriously considered again for decades. Then, starting in the 1990s, a series of discoveries suggested that the "dark energy" represented by the cosmological constant does indeed exist. Current measurements indicate that the density of this dark energy throughout the universe is much greater than the density of matter—both luminous matter and dark matter combined.

Though astronomers have measured the presence of this dark energy, we still have no idea what causes this energy, nor do we have a clue what this energy is made of. The quest to understand the cosmological constant in general, and dark energy in particular, is one of the great unsolved questions in astronomy today.

## What is **dark matter made of**?

Nobody has any real idea of what dark matter is. There exist some educated guesses, such as a new class of "weakly interacting massive particles" (WIMPs) or huge agglomerations of them ("WIMPzillas"); another class of "charged undifferentiated massive particles" (CHUMPs); or very light, neutral subatomic particles called neutralinos. No dark matter particle has ever been detected, however, so these possibilities are still nothing more than educated guesses.

## How does **dark matter affect** the **shape** of the **universe**?

Dark matter in the universe exerts a gravitational pull in the expanding universe. The more dark matter there is in the universe, the more likely it would be that the universe would have a closed geometry, and that the universe would end in a Big Crunch.

## How does **dark energy affect** the **shape** of the **universe**?

Dark energy apparently counteracts gravity by making space expand more energetically. If the amount of dark energy in the universe is, as astronomers think, proportional to the amount of space, then the continuing expansion of the universe means that the total amount of dark energy keeps increasing. Since the total amount of mass in the universe is not increasing, that means that the expansive effect of dark energy will ultimately overcome the contractive effect of dark matter. The more dark energy there is, the more open the geometry of the universe will tend to be, and the faster the expansion rate of the universe will increase over time.

## Have astronomers determined the matter and energy density of the universe?

**B**ased on measurements of the gravitational effects of dark matter and luminous matter in the distant universe, astronomers have measured $\Omega$ (that is, $\Omega_{DM} + \Omega_B$) to be about 0.3. Meanwhile, based on detailed observations of distant Cepheid variables and Type Ia supernovae, astronomers have deduced that the expansion rate of the universe is increasing. That means that $\Lambda$ is greater than zero. Finally, based on careful study of the cosmic microwave background, astronomers have confirmed that the universe has a flat geometry, meaning that $\Omega + \Lambda = 1$. Carrying the precision of these measurements to two decimal places, the current measurements show that $\Omega = 0.27$ and $\Lambda = 0.73$. If these numbers hold true, then our universe is destined to expand forever and there will be no Big Crunch.

## How do astronomers describe the **concentration of matter** in the **universe**?

Astronomers use the Greek capital letter omega ($\Omega$) to represent the concentration, or density, of matter in the universe. Sometimes, a subscript M is added ($\Omega_M$) to make clear that this is the concentration of matter; at other times, two subscripts are used to distinguish the concentration of dark matter ($\Omega_{DM}$) and of "baryonic" or non-dark matter ($\Omega_B$).

If dark energy does not exist, then the matter density in the universe alone determines the geometry and final fate of the cosmos. In that case, there are three possibilities. If $\Omega$ is larger than one, then the universe would have a closed geometry and ultimately collapse in a Big Crunch. If it is equal to one, then the universe would have a flat geometry and would expand forever. If it is less than one, then the universe would have an open geometry and would also expand forever.

## How do astronomers describe the **concentration of dark energy** in the universe?

Astronomers use the Greek capital letter lambda ($\Lambda$) to represent the concentration, or density, of dark energy in the universe. Sometimes, because dark energy also affects the geometry of the universe, the dark energy density is represented by a subscripted component of omega ($\Omega_\Lambda$).

If dark energy does indeed exist, then the combined effect of the matter and energy density in the universe determines the geometry of the universe. Thus, if ($\Omega + \Lambda$) or, equivalently, ($\Omega_\Lambda + \Omega_M$) is smaller than one, then the universe is open; if it is greater than one, then the universe is closed; and if it is equal to one, then the universe has flat geometry.

53

# MULTI-DIMENSION THEORIES

### What caused the **hyperinflation** of the **universe**?

Nobody knows why hyperinflation occurred in the early universe. One possibility is that, as the universe aged and cooled down, the fundamental forces in the universe began splitting apart from one another, and one of those splits released titanic amounts of energy that powered inflation.

### What is **spontaneous symmetry breaking**?

Spontaneous symmetry breaking is a physical phenomenon by which something balanced becomes permanently unbalanced. One example might be a ball sitting on top of a hill: it is exactly balanced, but if the ball suddenly rolls down the hill to the bottom, then the system is no longer balanced; since the ball will not roll uphill by itself, the system is now permanently unbalanced. Most of us can think of symmetry from, say, the example of paper folding. From a more general point of view, symmetry can be viewed as a measure of the order or complexity of a system: for example, a crystal.

Theoretical cosmologists hypothesize that the fundamental forces of the universe split off from one another in a form of spontaneous symmetry breaking. If a single, unified force existed with a certain "symmetry" just after the Big Bang, and if that symmetry were somehow "broken" so that the unified force were fractured, then the result might be several fundamental forces. Also, there would be a huge release of stored-up energy that might power hyperinflationary expansion or other kinds of activity in the early universe.

### What is **supersymmetry**?

Supersymmetry is part of a hypothetical model of how the universe works. It gives one explanation of how the universe might have evolved into its current state and

## How can 10 or 11 dimensions exist in the universe?

One hypothetical model that explains how so many dimensions can exist in our universe is the idea of compactification. Picture a big oil or natural gas pipeline stretching across a vast plain: when one stands next to it, it clearly has the three dimensions of length, depth, and height. By moving away a few feet, it increasingly seems like the pipeline has only length and height; moving still farther away, it may seem like it only has one dimension: length. In a sense, two of the pipeline's three dimensions have now been "compactified." They are still there, but they are too small to be observed. The same concept might apply to dimensions beyond those of observable space and time. This idea has been around for decades. However, it might be impossible to confirm the existance of other dimensions because scientists would have to observe size scales smaller than the Planck length to see the amount of compactification necessary for the universe to behave as it does.

suggests that the universe is unified under a single symmetric framework. One prediction of the supersymmetric model is that, for every kind of fundamental particle in the universe, there are symmetric partner particles or "sparticles" that exist as well, but these are not readily observable. So far, no sparticle has yet been detected. For this and other reasons, supersymmetry in the universe has not yet been confirmed.

## What is the **"Grand Unified Theory"**?

Some scientists think that all the physical laws of the universe might be described by a single theory. One well-known scientist to study this "Grand Unified Theory" idea was none other than Albert Einstein. He failed to create such a theory, but he laid the groundwork for others who have been plugging away at it ever since. Many of these models for a theory of everything are promising, but they are very complex and still in their scientific infancy.

## What is the **best known "Theory of Everything"** that is currently being examined?

String theory is the name of the best-known hypothetical model that tries to unify all the behavior of everything in the universe. The basic idea is that the particles in our universe are only four-dimensional parts of multi-dimensional structures called strings. In this model, when particles interact in our universe, they are really interacting on many dimensions. The results, even though they may appear to be completely new particles, are just different "vibrational modes" of the same strings.

## How can scientists prove that any of these theories are true?

That is a huge obstacle in theoretical cosmology today. Scientists have proposed some experiments that might confirm some of the predictions of these cosmological hypotheses, but the technology required to conduct them successfully does not yet exist. For example, one formulation of membrane theory suggests that when a very massive star explodes in a supernova, a tiny fraction of a percent of its energy might escape the universe and "leak" onto one of the membranes. But supernovae are rare—one occurs in our Milky Way only every century or so—and our modern telescopes and instruments cannot possibly measure the total energy released by a supernova with nearly enough precision.

## According to **string theory**, how many **dimensions** are there in the universe?

Currently, the most accepted model of string theory contains eleven dimensions. This 11-dimensional "supersymmetric bulk" can spawn or anchor 10-dimensional strings, which interact and create a 4-dimensional result that is our universe.

## What is a 'brane?

A membrane, or 'brane for short, is a multi-dimensional structure that perhaps exists in something like a supersymmetric bulk described above. Membranes can move around in the bulk—like a huge jellyfish floating and swirling around in a vast ocean—and interact with (that is, "bump into") one another, possibly causing vast releases or exchanges of energy. A cosmological model that invokes 'branes is often called a membrane theory or M theory. There are even different kinds of 'branes, which are given names like m-'branes, n-'branes, or p-'branes.

## According to **membrane theory**, **where** is the **universe located**?

Hypothetically, it is possible to think of two five-dimensional membranes that contact one another in one or more dimensions. The multidimensional intersection of those 'branes could be, for example, a point, a line, a surface, or even a four-dimensional spacetime. One idea, then, is that our universe, which is a four-dimensional spacetime, exists because two membranes intersected, launching the expansion of space and the beginning of time. That moment of intersection would thus be the Big Bang, and the universe would then be at the intersection of two 'branes.

## How might **studying the smallest particles** help resolve the **origin of the universe**?

One important area in the study of the Big Bang and the origin of the universe is particle physics. By generating and studying the smallest, most energetic subatomic particles in huge particle accelerators, physicists can glimpse, if briefly, the conditions that might have existed in the early universe. This can be done, for example, by crashing atomic nuclei together that are moving at 99 percent or more of the speed of light, and then seeing what comes out of the wreckage.

# THE END OF THE UNIVERSE

### Why does it matter what **shape the universe** is?

The shape of the universe affects what the final fate of the universe will be. The universe is currently expanding, and if the geometry of the universe is closed, then the expansion would most likely slow gradually, then eventually stop, reverse into a contraction, and result in a "Big Crunch" in which the universe ends as a supertiny, superhot point like a Big Bang in reverse. If the geometry of the universe is flat or open, then the universe will most likely expand forever.

### What is the **current prediction** of our **universe's fate**?

Current observations show that the universe has flat geometry, and that there is a lot of dark energy in the universe. In fact, since $\Lambda = 0.73$, that means that 73 percent of the "stuff" in the universe is dark energy, which keeps getting more and more plentiful as the universe continues to expand. So the expansion rate of the universe is increasing, and we live in an accelerating universe.

### Will the **universe** ever **end**?

By "end," if one means that time will stop and the cosmos will cease to exist, then no, the universe will not end. After a very long time, the universe will reach a stage where literally nothing will happen. All the matter will be formless and disorderly, and all the energy will be so sparsely distributed that there will be no significant interactions of any kind, subatomic or otherwise. So, in a sense, that is also an "end" to the universe—not a fiery or definitive end, but an infinitely long period of cold, dark nothingness.

### What do scientists believe will **ultimately happen** to matter and energy in the **universe**?

The acceleration of the expansion of the universe is carrying all matter ever farther apart on cosmic scales. Eventually, gravity will not be able to overcome that expansion to form new, large structures. Some calculations even suggest that within a few billion years distant galaxies will no longer be observable to us. Then, all the stars in the universe will consume their raw materials and burn out, leaving stellar corpses throughout the cosmos. Those corpses—mostly white dwarfs and neutron stars—and all the other baryonic matter in the universe will then, if current ideas of particle physics are correct, undergo proton decay and disintegrate. Finally, the black holes in the universe will emit Hawking radiation until they evaporate completely. All that will be left will then be dark matter, dark energy, and a whole lot of disordered subatomic particles that pretty much do nothing.

### When will the universe ultimately "die"?

If current theories are correct, then all the stars will burn out within a million trillion years; all protons will decay within a million trillion trillion trillion years; and all black holes—even the supermassive ones—will evaporate within a million trillion trillion trillion trillion trillion trillion trillion years. The universe is predicted to die, in other words, in about a googol ($10^{100}$) years, give or take a factor of a hundred.

# GALAXIES

## FUNDAMENTALS

### What is a **galaxy**?

A galaxy is a vast collection of stars, gas, dust, and dark matter that forms a cohesive gravitational unit in the universe. In a way, galaxies are to the universe what cells are to the human body: each galaxy has its own identity, and it ages and evolves on its own, but it also interacts with other galaxies in the cosmos. There are many, many different kinds of galaxies; Earth's galaxy is called the Milky Way.

### How many galaxies are there in the universe?

Thanks to the finite speed of light and the finite age of the universe, we can only see the universe out to a boundary called the cosmic horizon, which is about 13.7 billion light-years in every direction. Within this observable universe alone, there exist an estimated 50 to 100 billion galaxies.

### What **kinds of galaxies** are there?

Galaxies are generally grouped by their appearance into three types: spiral, elliptical, and irregular. These groups are further subdivided into categories like barred spiral and grand design spiral, giant elliptical and dwarf spheroidal, and Magellanic irregular or peculiar.

Galaxies are also often categorized by characteristics other than their appearance. For example, there are starburst galaxies, merging galaxies, active galaxies, radio galaxies, and many more.

Spiral galaxy NGC 4414. (*NASA, The Hubble Heritage Team, STScI, AURA*)

## How are **galaxies classified** or organized?

In the 1920s the pioneering astronomer Edwin Hubble, who devoted his life to studying galaxies, proposed a way to classify galaxies based on their shapes. He proposed a "sequence" of galaxy types: from E0 (sphere-shaped elliptical galaxies) to E7 (cigar-shaped ellipticals), S0 (lenticular galaxies) to Sa and SBa (spiral galaxies with large bulges and bars), Sb and SBb (spirals with medium-sized bulges and bars), and Sc and SBc (spirals with small bulges and bars). The sequence is known as the Hubble sequence, and it is often shown visually as a Hubble "tuning fork diagram."

## What is an **elliptical galaxy**?

An elliptical galaxy is a galaxy that appears to be an ellipse from our point of view. The ellipticity of the galaxy—how round or flat it is—varies greatly in ellipticals, so they can look like anything from perfect spheres to long cigars. Based on observations and theoretical models, astronomers think that the three-dimensional shape of elliptical galaxies are triaxial; that is, they each can have a length, width, and height that are all different from one another. So ellipticals can be shaped like gigantic basketballs, rugby balls, ostrich eggs, cough drops, Tic-Tacs, and anywhere in between.

## What is a **spiral galaxy**?

A spiral galaxy is a galaxy that appears to have spiral-shaped structures, or arms, that contain bright stars. As it turns out, those spiral arms are not solid structures, but rather spiral density waves: ridges in a disk of spinning dust and gas that are denser than the regions surrounding them. Spiral galaxies have star-filled, ellipsoid-shaped bulges at their centers; a star-filled, thin disk of spinning gas surrounding the bulge; and a sparsely populated stellar halo that envelops both the disk and the bulge.

## What is a **barred spiral galaxy**?

Some spiral galaxies have spiral arms that emanate not at the center of the galaxy but at some distance away from the center. The bulges of these galaxies are actually elongated, bar-shaped structures that contain billions of stars. These kinds of spirals are called barred spiral galaxies.

## What is a **lenticular galaxy**?

A lenticular galaxy is a lens-shaped galaxy that has strong elements of both elliptical and spiral galaxies. It can be viewed as either an elliptical galaxy with a disk surrounding its outer edge, or as a spiral galaxy with a very large bulge and almost no spiral arm structure. A very striking visual example of a lenticular galaxy is the object called Messier 104, which is nicknamed "The Sombrero."

## What is an **irregular galaxy**?

An irregular galaxy is a galaxy that does not fit well into the standard categories of elliptical, spiral, or barred spiral galaxies. Two examples of irregular galaxies are the Large Magellanic Cloud and Small Magellanic Cloud, which are visible from Earth's southern hemisphere. Irregular galaxies can have some spiral or elliptical structure, while also having other kinds of components, such as wispy trails of stars and gas.

## What is a **peculiar galaxy**?

A peculiar galaxy is a galaxy that could probably fit into one of the other general categories of galaxies—elliptical or spiral—except for some kind of peculiarity. Some examples include a long tail of stars, an unusually shaped disk, a second bulge, or even a whole other galaxy overlapping it or crashing into it. Many

NGC 4650A, an example of a peculiar galaxy, is the result of two galaxies crashing into one another. In this amazing example of such an instance, the result has been described as a "polar ring" galaxy because it appears as if one galaxy is traveling through a ring formed by the second. (*NASA, The Hubble Heritage Team, STScI, AURA*)

61

The motion of the stars and gas in the galaxies determines how galaxies maintain their shapes. In elliptical galaxies, the stars must move in random directions in all kinds of differently shaped orbits, like a swarm of bees buzzing around a central point. In spiral galaxies, the stars must revolve in nearly circular, stable orbits around a single point and are in the shape of a thin, orderly disk. If that orderly revolution is disrupted—for example, by a collision from another galaxy—then the disk shape will be disrupted as well. It could be permanently destroyed to be replaced by a chaotic ellipsoidal swarm.

peculiar galaxies look the way they do because they are in the process of colliding, interacting, or merging with other galaxies.

## Why do **galaxies** have **different shapes**?

Originally, when Edwin Hubble created his tuning fork diagram, he hypothesized that galaxies followed a sequence as they aged: all galaxies would be elliptical at first, and then flatten out over time as they spun. This idea was disproven, however.

Modern computer simulations and mathematical calculations now show that all galaxies form by the collection of smaller clumps of matter—subgalactic "clumps" that fall together into a single gravitational unit. If lots of little clumps collect, it usually becomes a spiral galaxy; but if there are some very large clumps that come together toward the end of the process, then it usually results in an elliptical galaxy. This model of galaxy shape-forming appears to be generally correct, but a number of details still need to be worked out to make a coherent theory.

## How **big** are **galaxies**?

Galaxies range greatly in size and mass. The smallest galaxies contain perhaps 10 to 100 million stars, whereas the largest galaxies contain trillions of stars. There are many more small galaxies than large ones. The Milky Way, which has at least 100 billion stars, is on the large end of the scale; its disk is about 100,000 light-years across.

## What is a **dwarf galaxy**?

Dwarf galaxies, as the name suggests, are galaxies that have the least mass and fewest stars. The Large Magellanic Cloud, a galaxy that orbits the Milky Way, is considered a large dwarf galaxy; it contains, at most, about one billion stars. Like larger galaxies, dwarf galaxies come in many forms, including dwarf ellipticals, dwarf spheroidals, and dwarf irregulars.

Massive galaxies may have formed while the universe was young when galaxy collisions were more common. In this artist's rendering, a super-massive galaxy with high amounts of dust emits radio jets from a central black hole. (*NASA/JPL-Caltech/T. Pyle*)

## How are **galaxies distributed** throughout the universe?

Observations show that galaxies are distributed unevenly throughout the universe. Instead of all galaxies being about the same distance apart, the majority of galaxies are collected along vast filamentary and sheetlike structures many millions of light-years long. These filaments and sheets connect at dense nodes—clusters and super-clusters—of galaxies, and the net result is a three-dimensional weblike distribution of matter in the universe nicknamed the Cosmic Web. Between the filaments and sheets are large pockets of space with relatively few galaxies; these sparse regions are called voids.

## What is a **group of galaxies**?

A group of galaxies usually contains two or more galaxies the size of the Milky Way or larger, plus a dozen or more smaller galaxies. The Milky Way and Andromeda galaxies are the two large galaxies in the Local Group. There are a few dozen smaller galaxies in the group, including the Magellanic Clouds, the dwarf elliptical Messier 32, the small spiral galaxy Messier 33, and many small dwarf galaxies. The Local Group of galaxies is a few million light-years across.

63

## What is a **cluster of galaxies**?

A cluster of galaxies is a large collection of galaxies in a single gravitational field.
Rich clusters of galaxies usually contain at least a dozen large galaxies as massive
as the Milky Way, along with hundreds of smaller galaxies. At the center of large
clusters of galaxies there is usually a group of elliptical galaxies called "cD" galaxies. Clusters of galaxies are usually about ten million light-years across. The Milky
Way galaxy is near, but not in, the Virgo cluster, which itself is near the center of
the Virgo supercluster.

## What is a **supercluster of galaxies**?

Superclusters of galaxies are the largest collections of massive structures. They
occur at the nodes of large numbers of matter filaments in the cosmic web, and are
up to a hundred million or more light-years across. There are usually many clusters of galaxies in a supercluster, or a single very large cluster at its center, along
with many other groups and collections of galaxies that are collected in the supercluster's central gravitational field. Superclusters contain many thousands—and
sometimes millions—of galaxies. The Milky Way galaxy is located on the outskirts
of the Virgo supercluster.

## What is a **field galaxy**?

A field galaxy usually refers to a galaxy that has few or no neighboring galaxies.
Many field galaxies are actually members of small groups of galaxies, but no field
galaxies are parts of rich clusters of galaxies. The vast majority—about 90 percent—
of all galaxies in the universe are considered by astronomers to be field galaxies.

## What kinds of **galaxies** are the **most common**?

That depends on the galaxies' environment. Among field galaxies and group galaxies, spiral galaxies are much more common than elliptical galaxies. In rich clusters,

however, the opposite is true. Interestingly, the further back in the history of the universe we look, the more common irregular and peculiar galaxies are. At all times in cosmic history, there have always been many more faint, dwarf galaxies than luminous, large galaxies like the Milky Way.

## What are **some well-known galaxies**?

The table below lists some of the galaxies that are well known to both professional and amateur astronomers.

### Some Well-Known Galaxies

| Common Name | Catalog Name | Galaxy Type |
| --- | --- | --- |
| Andromeda Galaxy | Messier 31 | spiral |
| Antennae | NGC 4038/4039 | interacting |
| Cartwheel Galaxy | ESO 350-40 | spiral ring |
| Centaurus A | NGC 5128 | elliptical |
| Flagellan | G515 | peculiar elliptical |
| Messier 49 | NGC 4472 | elliptical |
| Messier 61 | NGC 4303 | barred spiral |
| Messier 87 | NGC 4486 | elliptical |
| Mice | NGC 4676 | interacting |
| NGC 1300 | ESO 547-31 | barred spiral |
| Pinwheel Galaxy | Messier 101 | spiral |
| Sombrero Galaxy | Messier 104 | lenticular |
| Southern Pinwheel | Messier 83 | spiral |
| Triangulum Galaxy | Messier 33 | spiral |
| Whirlpool Galaxy | Messier 51 | spiral |

# THE MILKY WAY

## What is the **Milky Way galaxy**?

The Milky Way is the galaxy we live in. It contains the Sun and at least one hundred billion other stars. Some modern measurements suggest there may be up to 500 billion

When viewed at night from a location free of clouds or light pollution, the Milky Way resembles a milky spray of light across the sky. (*iStock*)

stars in the galaxy. The Milky Way also contains more than a billion solar masses' worth of free-floating clouds of interstellar gas sprinkled with dust, and several hundred star clusters that contain anywhere from a few hundred to a few million stars each.

## What **kind of galaxy** is **the Milky Way**?

Figuring out the shape of the Milky Way is, for us, somewhat like a fish trying to figure out the shape of the ocean. Based on careful observations and calculations, though, it appears that the Milky Way is a barred spiral galaxy, probably classified as a SBb or SBc on the Hubble tuning fork diagram.

## Where is the **Milky Way** in our **universe**?

The Milky Way sits on the outskirts of the Virgo supercluster. (The center of the Virgo cluster, the largest concentrated collection of matter in the supercluster, is about 50 million light-years away.) In a larger sense, the Milky Way is at the center of the observable universe. This is of course nothing special, since, on the largest size scales, every point in space is expanding away from every other point; every object in the cosmos is at the center of its own observable universe.

## Within the Milky Way galaxy, **where is Earth** located?

Earth orbits the Sun, which is situated in the Orion Arm, one of the Milky Way's spiral arms. (Even though the spiral arms of the Milky Way or any other galaxy are

## Why is our galaxy called the Milky Way?

**B**arred spiral galaxies are comprised of a disk that has the vast majority of stars in the galaxy, and a bar-shaped bulge at the center that also contains a large concentration of stars. In Earth's night sky, the disk of the galaxy stretches all the way around the sky and is about as wide as an outstretched hand. If one looks up at it with the unaided eye, it appears as a starry stream of light stretching from one side of the sky to the other. Ancient Chinese astronomers called this band of light the "Silver River," while ancient Greek and Roman astronomers called it a "Road of Milk" (*Via Lactea*). This was translated into English as the "Milky Way." When astronomers realized that we live in a galaxy, the name Milky Way was used to refer not just to this band of stars, but also to the entire galaxy.

not solid structures, the size scale of the galaxy is so large that the density wave will last for millions of years; it is therefore appropriate to say we are "in" the arm at this period in cosmic history.) Earth and the Sun are about 25,000 light-years away from the galactic center.

## How **large** is the **Milky Way**?

Current measurements indicate that the stellar disk of the Milky Way is about 100,000 light-years across and 1,000 light-years thick. If the Milky Way disk were the size of a large pizza, then the solar system might be a microscopic speck of oregano halfway out from the center to the edge of the crust. The bar-bulge structure of the Milky Way is about 3,000 light-years high and maybe 10,000 light-years long.

If you take into account the dark matter in the Milky Way, its size increases dramatically. Based on current measurements, at least 90 percent of the mass in the Milky Way's gravitational field is made up of dark matter, so the luminous stars, gas, and dust of the galaxy are embedded at the center of a huge, roughly spherical dark matter halo more than a million light-years across.

## How **fast is Earth moving** within the **Milky Way** galaxy?

Earth (and the solar system) is moving through the Milky Way's disk in a

An artist's conception of our barred spiral galaxy and the location of the Sun within it. (*NASA/JPL-Caltech/R. Hurt*)

stable, roughly circular orbit around the galactic center. The latest astronomical measurements indicate that our orbital velocity around the center of the Milky Way is about 450,000 miles per hour (200 kilometers per second). That is almost a thousand times the cruising speed of most commercial jetliners. Even so, the Milky Way is so huge that one complete orbit takes about 250 million years.

### What were some of the **earliest studies** of the **Milky Way**?

In the early 1600s, Galileo Galilei first examined the Milky Way through a telescope and saw that the glowing band of light was made up of a huge number of faint stars that were apparently very close together. As early as 1755, the German philosopher Immanuel Kant suggested that the Milky Way is a lens-shaped collection of stars and that there may be many such collections in the universe. The German-English astronomer William Herschel (1738–1822), who is perhaps best known for his discovery of the planet Uranus, was also the first astronomer to conduct a scientific survey of the Milky Way.

### What is the **warp** in the **Milky Way** galaxy?

Unfortunately, unlike certain popular science fiction shows, the "warp factor" in the Milky Way is not a way to travel faster than the speed of light. The disk of the Milky Way galaxy is actually not perfectly flat. Aside from its slight thickness, it is also somewhat warped in sort of the same way a spinning pizza crust tossed into the air warps and wobbles as it rotates. Of course, since our galaxy is far bigger than a pizza, the warp takes millions of years to make its way around the disk even once.

Astronomers think that the gravitational effects of one or more dwarf galaxies falling into the much larger Milky Way caused the warp. Such a relatively small impact would not destroy the disk structure of our galaxy but could have caused the disk to buckle a little bit.

# THE MILKY WAY'S NEIGHBORHOOD

## What **other galaxies** are **near** the **Milky Way** galaxy?

"Near" is a relative term when it comes to galaxies. Within a few million light-years of the Milky Way are several dozen galaxies that make up the Local Group. Some of those galaxies, such as the Sagittarius dwarf galaxy, are almost in physical contact with the Milky Way's outskirts.

## What is the **largest galaxy** in the **Local Group**?

The Andromeda galaxy, which is slightly larger than the Milky Way, is the largest galaxy in the Local Group. Andromeda is also known as Messier 31, or M31, because it is the thirty-first object listed in the famous catalog of night-sky objects compiled by Charles Messier in 1774.

## When was **Andromeda discovered**?

On a perfectly clear, moonless night, the Andromeda galaxy can just barely be seen by the naked eye. So it is likely that ancient astronomers knew of its existence, but did not understand what it was. According to French astronomer Charles Messier, who put the great nebula in Andromeda as the thirty-first object in his famous Messier catalog, the first European astronomer who discovered Andromeda was Simon Marius. Marius observed the Andromeda galaxy through a telescope in 1612; he was probably the first person to do so. According to non-European records, however, the ancient Persian astronomer Al-Sufi observed the Andromeda galaxy as early as 905 C.E. without the aid of a telescope. Al-Sufi called it the "little cloud."

Views of the Andromeda galaxy using both infrared and visible light. (*NASA/JPL-Caltech/K. Gordon*)

## How similar is the Andromeda galaxy to our own?

The Andromeda galaxy shares many characteristics with the Milky Way. It is a large spiral galaxy, like the Milky Way; it appears to be roughly the same age as the Milky Way; and it contains many of the same types of objects as the Milky Way, including a supermassive black hole at its center. Andromeda is somewhat larger than the Milky Way, but it is still close to 100,000 light-years (or about 600,000 trillion miles, or one million trillion kilometers) across.

## What **other galaxies** populate the **Local Group** of galaxies?

The three dozen or so galaxies in the Local Group other than Andromeda and the Milky Way are all dwarf galaxies. They vary in size from about one-half to one-thousandth the diameter of Andromeda and the Milky Way; they contain only a few million to a few billion stars each (compared to Andromeda and the Milky Way, which contain hundreds of billions of stars). The largest of these Local Group dwarf galaxies are the Large Magellanic Cloud and the Small Magellanic Cloud, which orbit the Milky Way, and Messier 32 and Messier 33, which orbit Andromeda. Other well-known Local Group dwarfs include IC 10, NGC 205, NGC 6822, and the Sagittarius dwarf galaxy. The table below lists some of the Local Group galaxies.

### Local Group Galaxies

| Galaxy Name | Galaxy Type | Distance (kiloparsecs) | Absolute Visual Magnitude |
|---|---|---|---|
| Milky Way | barred spiral | 0 | −20.6 |
| Sagittarius | dwarf elliptical | 24 | −14.0 |
| Large Magellanic Cloud | irregular | 49 | −18.1 |
| Small Magellanic Cloud | irregular | 58 | −16.2 |
| Ursa Minor | dwarf elliptical | 69 | −8.9 |
| Draco | dwarf elliptical | 76 | −8.6 |
| Sculptor | dwarf elliptical | 78 | −10.7 |
| Carina | dwarf elliptical | 87 | −9.2 |
| Sextans | dwarf elliptical | 90 | −10.0 |
| Fornax | dwarf elliptical | 131 | −13.0 |
| Leo II | dwarf elliptical | 230 | −10.2 |
| Leo I | dwarf elliptical | 251 | −12.0 |
| Phoenix | irregular | 390 | −9.9 |
| NGC 6822 | irregular | 540 | −16.4 |
| NGC 185 | elliptical | 620 | −15.3 |
| IC 10 | irregular | 660 | −17.6 |
| Andromeda II | dwarf elliptical | 680 | −11.7 |
| Leo A | irregular | 692 | −11.7 |
| IC 1613 | irregular | 715 | −14.9 |

## What important astronomical event occurred recently in the Large Magellanic Cloud?

On February 23, 1987, Supernova 1987A appeared in the Large Magellanic Cloud. It was discovered almost immediately by two astronomers, Ian Shelton and Oscar Duhalde, at Las Campañas Observatory in Chile. This event was significant to astronomers because it was the closest supernova—a titanic stellar explosion—to have been observed in hundreds of years. The event has given astronomers one of the most valuable stellar laboratories ever to examine how stars are born, live, and die. Supernova 1987A is still being carefully studied today.

| Galaxy Name | Galaxy Type | Distance (kiloparsecs) | Absolute Visual Magnitude |
|---|---|---|---|
| NGC 147 | elliptical | 755 | −14.8 |
| Pegasus | irregular | 760 | −12.7 |
| Andromeda III | dwarf elliptical | 760 | −10.2 |
| Andromeda VII | dwarf elliptical | 760 | −12.0 |
| Messier 32 | elliptical | 770 | −16.4 |
| Andromeda | spiral | 770 | −21.1 |
| Andromeda IX | dwarf elliptical | 780 | −8.3 |
| Andromeda I | dwarf elliptical | 790 | −11.7 |
| Cetus | dwarf elliptical | 800 | −10.1 |
| LGS 3 | irregular | 810 | −9.7 |
| Andromeda V | dwarf elliptical | 810 | −9.1 |
| Andromeda VI | dwarf elliptical | 815 | −11.3 |
| NGC 205 | elliptical | 830 | −16.3 |
| Triangulum | spiral | 850 | −18.9 |
| Tucana | dwarf elliptical | 900 | −9.6 |
| WLM | irregular | 940 | −14.0 |
| Aquarius | irregular | 950 | −10.9 |
| Sagittarius DIG | dwarf irregular | 1,150 | −11.0 |
| Antlia | dwarf elliptical | 1,150 | −10.7 |
| NGC 3109 | irregular | 1,260 | −15.8 |
| Sextans B | irregular | 1,300 | −14.3 |
| Sextans A | irregular | 1,450 | −14.4 |

## What is the **Large Magellanic Cloud**?

The Large Magellanic Cloud, or LMC, is the largest dwarf galaxy that orbits our own Milky Way galaxy. It is an irregular disk galaxy that is similar in shape to the Milky Way, and we see it sort of edge on, so it looks like an oblong-shaped cigar to viewers on Earth. The LMC is about 30,000 light-years across and 170,000 light-years

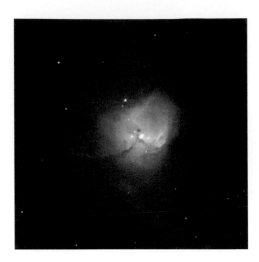

Located in the Small Magellanic Cloud, N81 is a cluster of about 50 stars within a mere 10 light-year distance of one another. Such unusual phenomena within both the Large and Small Magellanic Clouds make them irregular galaxies. ( *NASA, ESA, Mohammad Heydari-Malayeri Paris Observatory France* )

away from Earth. It is named after the explorer Ferdinand Magellan, who in 1519 was the first European to record its existence.

## What is the **Small Magellanic Cloud**?

The Small Magellanic Cloud (SMC), like its bigger compatriot the Large Magellanic Cloud (LMC), is a small irregular galaxy that orbits the Milky Way galaxy. It is a roughly disk-shaped galaxy about 20,000 light-years across and about 200,000 light-years away. Like the LMC, the SMC is forming stars at a rate much faster than that of the Milky Way. It is thus an important target for astronomers who are studying the formation and aging of stars and galaxies.

### Who first determined that the **Small Magellanic Cloud** is a **separate galaxy**?

American astronomer Harlow Shapley (1885–1972) earned his doctoral degree at Princeton University in 1913, working with Henry Norris Russell (1877–1957), who was famous for the Hertzsprung-Russell diagram. Shapley and Russell studied eclipsing binary stars, systems of two stars orbiting around one another in such a way that one star would regularly block the other star from our view. Later, while working at the Mount Wilson Observatory in Pasadena, California, he studied other kinds of variable stars, including RR Lyrae and Cepheid variable stars, which could be used as "standard candles" to measure distances. With them, he measured the distances to many of the globular star clusters that orbit around the Milky Way. By mapping out the positions of the globular clusters, Shapley showed that the disk of our Milky Way galaxy was some 100,000 light-years across—much larger than had been previously thought—and that our sun and solar system was off to one side of the Milky Way, rather than at its center.

In 1921 Shapley became the director of the Harvard College Observatory. There, he began to study variable stars in the Large and Small Magellanic Clouds. In 1924 he used those variable stars as standard candles to show that the Small Magellanic Cloud was at least two hundred thousand light years away from Earth, and thus must be a small galaxy of its own, rather than part of the Milky Way.

### What were the **Shapley-Curtis debates** of 1920?

During the first two decades of the twentieth century, Harlow Shapley believed that the Milky Way was the only major galaxy in the universe. Other scientists, such as

## Why is the Small Magellanic Cloud important in the history of observational cosmology?

The American astronomer Henrietta Swan Leavitt (1868–1921) and the Danish astronomer Ejnar Hertzsprung (1873–1967) studied Cepheid variable stars in the Small Magellanic Cloud in 1913. That work led to the first period-luminosity relation calculation for Cepheid variables and their potential use as standard candles for determining distances beyond the Milky Way. A decade later, Edwin Hubble used their work to determine that Andromeda was far outside the Milky Way, leading to the birth of modern extragalactic astronomy.

Heber Curtis (1872–1942), thought that the "spiral nebulae" were in fact galaxies like our own. To bring light to this very important scientific question of the time, a series of scientific debates were held in Washington, D.C., in 1920 between Shapley and Curtis. Each person laid out the scientific issue in his own way and compared the evidence of one position versus the other. In the end, Harlow Shapley was wrong and Heber Curtis was right: the Milky Way is indeed one galaxy among billions in the universe. Even though Shapley was wrong, he still is considered a great scientist today.

# GALAXY MOVEMENT

## How do astronomers **measure distances** to **galaxies**?

The original measurement of the distance from Earth to the Andromeda galaxy, which was done by Edwin Hubble in the 1920s, has been refined over the past century. Today, except for specific distance measurements to test particular astronomical methods, most astronomers use the Hubble Law—the relationship between redshift and distance—to measure the distance to distant galaxies.

## How does the **Hubble Law work**?

Edwin Hubble showed that the farther away a galaxy is from the point of observation, the faster it moves away because of the expansion of the universe. The Hubble Law gives the basic conversion factor between the redshift and the distance. Using the current best measurement of the Hubble Constant (the expansion rate of the universe), and adjusting for the geometry of the universe, astronomers simply measure the redshift of any galaxy and then use the conversion factor to get the distance to that galaxy.

## What is the relation between the **redshift** and the **Doppler shift** when observing very **distant objects**?

As Vesto Slipher, Edwin Hubble, and other pioneering astronomers showed nearly a century ago, Doppler shifts in astronomy indicate the motions of objects toward

**N**ot exactly. When observing distant galaxies, the measured redshift can still be converted to the corresponding Doppler shift using the relativistic Doppler formula. However, at those very large distances, the measured redshift relates only very little to the motion of the galaxy through space; rather, it relates almost completely to cosmological expansion—sizes, distances, and ages.

or away from the observer. "Blueshift" is Doppler shift of objects moving toward an observer, while "redshift" is Doppler shift of objects moving away. Since the expanding universe carries galaxies faster and faster away as distances increase, the redshift gets higher and higher as well. Beyond a distance of about one billion light-years, the redshift gets so large that Einstein's special theory of relativity becomes a factor in the motion, and the usual formula converting redshift into Doppler shift no longer holds. In those cases, a more complicated equation called the relativistic Doppler formula must be used.

### How is **cosmological redshift calculated**?

Cosmological redshift is calculated by (1) figuring out how much the observed wavelength is shifted from the rest wavelength, and (2) expressing that shift as a ratio of the rest wavelength. Although it sounds complicated, it really is not. It turns out that this redshift number is very useful when deriving properties of distant galaxies, such as age and distance.

Here is an example for illustration. Say an astronomer is measuring the spectrum of a distant galaxy. If the unredshifted rest wavelength of a spectral feature is 100 nanometers, but for this galaxy the feature appears at 200 nanometers, then the measured redshift is one. If the feature appears at 300 nanometers, the redshift is two; if it is at 400 nanometers, the redshift is three; and so on.

### How does **redshift** relate to **age** as well as **distance** of galaxies?

Astronomers have deduced that the redshift of an object is not merely a representation of how fast it is moving away from us, but also how much the universe has expanded since the light we see from a distant object actually left that object. If an astronomer observes that light from a galaxy has a redshift of one, then that light left that galaxy when the universe was half its current diameter; if the redshift is two, then the universe was one-third its current diameter; if the redshift is three, then the universe was one-fourth its current diameter. This pattern continues all the way out to the edge of the observable universe: as the redshift approaches infinity, then the size of the universe approaches zero, which is the Big Bang. That means that redshift is a way to measure the cosmological age of any distant object one is observing. An astronomer can relate any fractional size of the universe with a certain number of years before the present day, and thus compute the age of the universe at the time the object is being observed.

### What is **look-back time**?

Light—that is, any kind of electromagnetic radiation—travels through space at more than 186,000 miles per second. That means that if we see an object 186,000 miles away, it took one second for the light from that object to reach us. This, in turn, means that we are actually seeing the object as it existed one second ago. This effect is called look-back time.

For astronomical distances, look-back time can be a significant effect. The look-back time for the Sun is eight minutes; the look-back time for the planet Jupiter is almost an hour; and for the Alpha Centauri star system, the look-back time is nearly four and a half years.

## AGE OF GALAXIES

### How does **look-back time** affect **observations of galaxies**?

Galaxies are really, really far away from Earth compared to planets or stars. So look-back times to galaxies can be a substantial fraction of the total age of the universe. Every light-year of distance creates a look-back time of one year. If a galaxy is five billion light-years away, then we are seeing the galaxy as it existed five billion years ago, which is before our planet was even formed.

### How do **astronomers** use **look-back time** to study the **universe**?

In a sense, look-back time is unfortunate because we can only guess how distant galaxies look today. But on the flip side, astronomers can use look-back time to study how the universe has aged and evolved since the distant past. That is because we directly observe how distant galaxies looked back in the past—we do not need to rely on fossils or subjective writings, as biologists and historians might. It is as if we took a picture of a town or city many years ago and then compared it with another photograph taken recently to see how it has changed. With this tool, astronomers can figure out how the universe has evolved and changed going back to a time almost as early as the Big Bang 13.7 billion years ago.

### How **far away** are the **farthest galaxies**?

The most distant galaxies measured to date are at redshifts between 6 and 7, which puts them between 12 billion and 13 billion light-years away. Since the distance to the cosmic horizon is 13.7 billion light-years, that means that astronomers have looked more than 90 percent of the distance out to the edge of the observable universe.

## Is there anything further away than the farthest galaxies?

**A**ccording to current astronomical theories, there may be objects even more distant. But because of the effect of look-back time, these distant objects are also the oldest objects, so they may either be too faint to be detected with modern telescopes, or they may have existed during a time when the universe was still not fully transparent.

The most distant objects found so far are galaxies. Until only a few years ago, the most distant objects known were quasi-stellar objects (QSOs), which we now know reside in galaxies anyway. These days, both QSOs and non-QSO galaxies regularly vie for the title of most distant known objects.

## How old are the **oldest galaxies**?

Because of the phenomenon of look-back time, the most distant galaxies yet observed are also the oldest galaxies yet observed. Those galaxies have redshifts between 6 and 7, indicating they are both almost 13 billion light-years away and almost 13 billion years old.

## When did **galaxies form**?

Since the most distant—and thus, earliest—known galaxies in the universe have confirmed redshifts of between 6 and 7, the first galaxies must have formed even earlier than that. Current models of galaxy formation indicate that the first galaxies were probably assembled between redshift 10 and 20, or a little more than 13 billion years ago.

## What is a **quasi-stellar object**, or QSO?

A quasi-stellar object (QSO) is the general term given to an "active galactic nucleus" (AGN) that has very high luminosity. QSOs are so named because in typical astronomical images taken in visible light they usually look like stars, or stars with a little bit of fuzz or structure surrounding them. In fact, they are not stars at all, but they are so luminous compared to their host galaxy that they drown out the light from it.

## Is the **Milky Way** galaxy an **old galaxy**?

The Milky Way galaxy is certainly old—at least 10 billion years old—but current studies show that the Milky Way is not among the very oldest galaxies, which formed more than 13 billion years ago.

# GALACTIC DUST AND CLOUDS

## What is the **interstellar medium**?

The interstellar medium is the matter that exists within galaxies, between and among—but not including—the stars. Almost all of the interstellar medium is comprised of gas and microscopic dust particles.

## How much **interstellar medium** is there in **galaxies**?

About one percent of the luminous mass of a galaxy like the Milky Way (that is, excluding the non-baryonic dark matter) is interstellar medium. The rest of the mass consists primarily of stars and the end stages of stellar evolution, including white dwarfs, neutron stars, and black holes.

## How **dense** is the **interstellar medium**?

On average, the interstellar medium in our region of the Milky Way galaxy has a density of about one atom of gas per cubic centimeter. By contrast, Earth's atmosphere at sea level contains about $10^{19}$ gas molecules per cubic centimeter. There is also about one dust particle per 10,000,000 cubic meters in the local interstellar medium.

In some places, the interstellar medium can be much denser. When there is a large enough concentration of gas and dust in a given place, the interstellar medium can form clouds that are thousands of times denser than one atom per cubic centimeter. Even so, these interstellar clouds are millions of times less dense than the best laboratory vacuum chambers can produce on Earth.

## What does the **interstellar medium look like**?

It can appear in an amazing variety of forms and colors. Much of the interstellar medium is invisible; in fact, it will block the view of distant astronomical objects. However, through various physical processes, the interstellar medium can collect in special configurations and produce beautiful nebulae with remarkable shapes and sizes. The names of some of these nebulae reveal some of their charm: the Rosette, the Cat's Eye, the Hourglass, the Clownface, and the Veil.

## What is a **molecular cloud**?

A molecular cloud is a cloud that contains molecules—constructs of multiple atoms. The fact that the clouds contain molecules is very interesting in its own right. What is even more interesting, though, is that if an interstellar cloud can contain molecules it means that the cloud also harbors the conditions for the birth of new stars.

## Are **molecules** in the **interstellar medium** only found in molecular clouds?

No. They exist in the interstellar environments surrounding stars, too. Gas molecules in space, however, are much more fragile than atomic gas. Ultraviolet radiation from stars, for example, readily destroys molecules, breaking them up into atoms again. So the density of dust in molecular clouds helps to shield the molecules floating inside them, and allows them to stay together for long periods of time.

## How **big** are **molecular clouds**?

Molecular clouds can be enormous compared to stars. The largest ones are called "giant molecular clouds" and can be many light-years across. Giant molecular clouds can contain thousands or even millions of times as much mass as the Sun; they may also contain a number of dense core regions, each with 100 to 1,000 Suns' worth of gas. This is the raw material needed to build entire clusters of new stars.

## Where can the **interstellar medium** be **found in galaxies**?

Elliptical galaxies generally do not have very much interstellar medium compared to spiral galaxies. The Milky Way, for example, is a spiral galaxy, and contains an amount of interstellar medium several billion times the mass of the Sun. An elliptical galaxy about the same size as the Milky Way, in comparison, would likely not have even half that amount. Irregular galaxies have the largest proportion of their mass as interstellar medium. In most galaxies, the majority of their interstellar gas and dust collects in the disks of the galaxies rather than in bulges or halos.

## How does the **interstellar medium** affect **astronomical observations**?

The interstellar medium is, of course, itself a target for astronomical study. However, it can also complicate astronomical observations substantially. Think about sunsets on Earth. For some reason, the Sun looks much redder than it does during the middle of the day. That is because when the Sun is low in the sky it shines through dustier air; the dust in the sky tends to absorb proportionately more blue light and allow proportionately more red light to shine through. This effect of dust is called extinction, and it both changes the observed colors of astronomical objects and obscures them from view.

## Is interstellar dust similar to household dust?

No. Interstellar dust is typically much smaller, and it is made of very different material, compared to house dust here on Earth. While house dust typically is made up of things like dirt, sand, cloth fibers, crumbs, animal and plant residue, and even microscopic living creatures, interstellar dust is composed primarily of carbon and silicate (silicon, oxygen, and metallic ions) material, which is sometimes mixed with frozen water, ammonia, and carbon dioxide.

## Why is the **interstellar medium important**?

Every large thing in the universe is made up of smaller components. To make things like stars, planets, plants, and people, enough of the interstellar medium has to come together and interact—physically, chemically, and even biologically—to create them. In other words, we here on Earth are part of the interstellar medium. So to understand our origins and ourselves, we must understand the interstellar medium, which includes the basic building blocks of everything we see in the cosmos.

# NEBULAE, QUASARS, AND BLAZARS

## What is a **nebula**?

A nebula, derived from the Latin meaning "mist," is any cloud or collection of interstellar medium in one location in space. Nebulae are produced in many different ways. For example, they can be gathered together by gravity, dispersed by stars, or lit up by a powerful radiation source nearby. As beautiful as nebulae are, however, most of them nonetheless contain only a few thousand atoms or molecules per cubic centimeter. This is many times sparser than even the best laboratory vacuum chambers on Earth can achieve.

## How many **kinds of nebulae** are there?

There are numerous kinds of nebulae, which bear informal as well as formal names. Generally, types of nebulae are described either by their appearance

The nebula NGC 604, located in the galaxy Messier 33, is 1,500 light-years across. (*NASA, Hui Yang University of Illinois*)

79

(for example, dark nebulae, reflection nebulae, and planetary nebulae) or the physical processes that create them (such as protostellar nebulae, protoplanetary nebulae, or supernova remnants).

## What are **dark nebulae**?

Dark nebulae are, well, dark. They look like black blobs in the sky. They are generally dark because they contain mainly cold, high-density, opaque gas, as well as enough dust to quench the light from stars behind them. One example of a dark nebula is the Coal Sack Nebula, located near the constellation Crux (The Southern Cross).

## What is a **reflection nebula**?

A reflection nebula is lit by bright, nearby light sources. The dust grains in them act like countless microscopic mirrors, which reflect light from stars or other energetic objects toward Earth. To the human eye, reflection nebulae usually look bluish. This is because blue light is more effectively reflected in this way than red light.

## What is an **emission nebula**?

An emission nebula is a glowing gas cloud with a strong source of radiation—usually a bright star—within or behind it. If the source gives off enough high-energy ultraviolet radiation, some of the gas is ionized, which means the electrons and nuclei of the gas molecules become separated and fly freely through the cloud. When the free electrons recombine with the free nuclei to become atoms again, the gas gives off light of specific colors. Which colors they emit depends on the temperature, density, and composition of the gas. The Orion Nebula, for example, glows mostly green and red.

## What are some of the **best-known gaseous nebulae**?

The table below lists some famous nebulae.

### Some Well-Known Gaseous Nebulae

| Common Name | Catalog Name | Nebula Type |
|---|---|---|
| Crab Nebula | Messier 1 | supernova remnant |
| Dumbbell Nebula | Messier 27 | planetary nebula |
| Eagle Nebula | Messier 16 | star forming region |
| Eskimo Nebula | NGC 2392 | planetary nebula |
| Eta Carina Nebula | NGC 3372 | star forming region |
| Helix Nebula | NGC 7293 | planetary nebula |
| Horsehead Nebula | Barnard 33 | dark nebula |
| Hourglass Nebula | MyCn 18 | planetary nebula |
| Lagoon Nebula | Messier 8 | star forming region |
| Orion Nebula | Messier 42 | star forming region |
| Owl Nebula | Messier 97 | planetary nebula |

| Common Name | Catalog Name | Nebula Type |
|---|---|---|
| Ring Nebula | Messier 57 | planetary nebula |
| The Coal Sack | N/A | dark nebula |
| Trifid Nebula | Messier 20 | star forming region |
| Veil Nebula | NGC 6992 | supernova remnant |
| Witch Head Nebula | IC 2118 | reflection nebula |

## What is a **quasar**?

The term "quasar" is short for "quasi-stellar radio source." The term came into general usage in the 1960s, when astronomers studying cosmic radio sources noticed that many of them looked like stars on photographs. Subsequent studies showed that they were not stars at all, but rather active galactic nuclei. Nowadays, the word "quasar" is often used to mean any quasi-stellar object (QSO), whether or not it emits radio waves.

## When and how were **quasars first found**?

In the 1950s and 1960s, astronomers in Cambridge, England, began to use the most sensitive radio telescopes of the day to map the entire sky. There have been several "Cambridge catalogs," each deeper and more detailed than the last. The common practice in modern astronomy is that, when an object is detected using one band of electromagnetic radiation, the same object is searched for in other bands as well to

An artist's concept of a quasar in a distant galaxy. (*NASA/JPL-Caltech/T. Pyle (SSC)*)

## What are blazars and BL Lacertae objects?

**B**L Lacertae was a radio source that, originally, was identified as a special kind of variable star. But after 3C 273 was shown to be a quasar, astronomers revisited the study of BL Lacertae and realized that it was a quasar, too. However, it was one that varied a great deal, and very unpredictably, in its brightness. Today, objects like BL Lacertae are called blazars. Their spectral characteristics are very different from quasars like 3C 273, and they emit a much higher fraction of their energy at gamma ray and X-ray wavelengths than most other QSOs. This phenomenon probably occurs because we see the central supermassive black hole at a different angle.

get a more comprehensive understanding of the object through all of its different types of light emission.

The third Cambridge (3C) catalog contains hundreds of radio sources, and astronomers took visible-light photographs of these sources to see what they would look like to our eyes. The 273rd object in the 3C catalog looked like a star. But when astronomers subsequently studied more carefully the light it emits, it was discovered that 3C 273 was actually an active galaxy far away from the Milky Way. In fact, 3C 273 was the first quasar ever discovered and identified as a distant "active galactic nucleus" (AGN).

### How were **quasars first identified** as distant, super-bright objects?

In 1962 the Dutch-American astronomer Maarten Schmidt (1929–), examining the spectrum of 3C 273, realized that its pattern of emission lines was very much like that of some Seyfert galaxies, but more extreme. Furthermore, those emission lines were shifted far toward the red wavelengths of the electromagnetic spectrum. As Edwin Hubble had shown, such a redshift signature indicated that the object was likely to be very far away in the universe. Using the redshift, Schmidt showed that 3C 273 was nearly two billion light-years away from Earth. Another calculation showed that the object was far more luminous than the Milky Way galaxy; including its radio emission, 3C 273 was emitting more light each second than the Sun would in more than a million years. Soon, other radio sources in the 3C catalog were shown to be quasi-stellar objects, quasars that were all at distances billions of light-years away from Earth.

### How **bright** can **quasars** (and **QSOs** in general) get?

The brightest quasars (and, in general, QSOs) are many thousands of times brighter than all the stars in our Milky Way galaxy put together.

### What does a **quasar** really **look like**?

Imagine a supermassive black hole that is millions or billions of miles across and is at the center of a rapidly spinning disk of superhot gas. Around the disk and the black hole is a thick, doughnut-shaped torus of thicker, cooler gas. Matter falling toward the black hole accumulates in the torus and slowly swirls into the gas disk on its way to the black hole. Finally, right near the black hole, two super-energetic jets of matter shoot outward, above and below the disk, with matter traveling at nearly the speed of light. These jets extend thousands, even millions of light-years out into space. That is the basic picture of a quasar, or quasi-stellar object (QSO).

# BLACK HOLES IN GALAXIES

### Aside from stars and the interstellar medium, **what else do galaxies contain**?

Galaxies often have large magnetic fields that run through and around their disks or bulges. Although at any particular location, the fields may be weak, the overall effect of those fields can be tremendous, affecting the motion of charged particles and interstellar medium throughout galaxies. Galaxies can also contain black holes.

### Does the **Milky Way** contain a **supermassive black hole**?

It most certainly does. The center of the Milky Way is in the direction of the constellation Sagittarius; right at the center, there is an object called Sag A* (pronounced "Sagittarius A-star") that emits much more X-rays and radio waves than expected for a star-sized body. After mapping the motions of stars near Sag A* for more than a decade, astronomers concluded that Sag A* is an invisible object that is more than three million times the mass of the Sun. The only kind of object like that in the universe is a supermassive black hole.

### Does every **galaxy** contain a **black hole**?

There are two general categories of black holes that have been observed: stellar black holes and supermassive black holes. Every galaxy that has contained very hot, luminous stars—stars 20 times or more the mass of the Sun—almost certainly contains stellar black holes.

### Does every **galaxy** contain a **supermassive black hole**?

No, but based on current observations, the majority of galaxies do contain one. Among nearby galaxies, more than 90 percent of all galaxies that have been measured so far appear to contain a supermassive black hole.

# ACTIVE GALAXIES

### What is an **"active galaxy,"** or an **"active galactic nucleus"** (AGN)?

If a supermassive black hole exists at the nucleus of a galaxy, it may accumulate matter from the stars and gas that surround it. If this matter is accumulated rapid-

83

ly—at a rate of a few Earth-masses per second or greater—tremendous amounts of energy can be generated as the matter falls toward the black hole. The energy that is released in this way can be much greater than that of the nuclear fusion of a star. In fact, such a supermassive black hole system can radiate more energy in a few seconds than our Sun can produce in thousands or even millions of years. These systems are called active galactic nuclei, or AGNs.

## Who **first discovered** and studied **active galaxies**?

The American astronomer Carl Seyfert (1911–1960) is credited with the discovery of active galaxies. Seyfert's general area of astronomical expertise was determining the spectroscopic properties, colors, and luminosities of stars and galaxies. In 1940, he went to work as a research fellow at the Mount Wilson Observatory in California, the same institution where Edwin Hubble made his most famous discoveries about galaxies. By 1943 Seyfert had discovered a number of spiral galaxies with exceptionally bright nuclei. These galaxies had unusual spectral signatures that had extremely strong and broad emission lines, indicating that very energetic activity was going on inside their nuclei. Today, those types of active galaxies are called Seyfert galaxies in his honor.

## How many **different kinds** of **active galaxies** are there?

AGNs can occur in any type or shape of galaxy—spiral, elliptical, or irregular. Depending on exactly how the energy radiates from the AGN, they can have very different appearances. This has led to a wide variety of types of AGN, such as Seyfert Type 1 and Seyfert Type 2 galaxies, radio galaxies, BL Lac objects, blazars, and radio-loud and radio-quiet quasars. Sometimes, to simplify matters, astronomers refer to very luminous AGN of all types as simply "quasars" or "quasi-stellar objects" (QSOs). AGNs are sometimes not particularly luminous, emitting substantially less light than the rest of their host galaxy. In these cases, they are called "low-luminosity AGNs" and, again, may have many different characteristics.

## What determines the **luminosity** of an **AGN**?

Sometimes the amount of intervening material—either in our galaxy, or in the AGN host galaxy—can diminish the amount of light we see from an active nucleus. That does not affect, however, the AGN's luminosity, or the total amount of energy that it emits. The single most important determinant of an AGN's luminosity is the rate at which matter falls toward its central supermassive black hole. Low-luminosity AGNs might have only a few Earth-masses' worth of material falling onto the central black hole per year. The most luminous AGNs, on the other hand, are accreting (gathering mass by infalling matter) at a rate equivalent to swallowing a million Earths per year.

## What are **radio galaxies**?

Radio galaxies are simply galaxies—usually very ordinary-looking elliptical galaxies when viewed by visible light—that radiate an unusually large amount of radio waves. Often, the total energy of the radio wave emission far exceeds that of the galaxy's visible light emission. The majority of the radio wave emission usually comes from huge, puffy "lobes" or narrow "jets" that can be much larger than the visible galaxy itself. The excess radio emission is probably produced when much of the energy generated by an AGN is carried away by highly energetic streams of matter, which then interact with the interstellar medium in and around the host galaxy and cause copious emissions of radio waves.

### What is the **unified model** of **active galactic nuclei**?

After decades of studying active galactic nuclei, astronomers have put together a single, unified model that might explain why all AGNs look the way they do. Basically, all AGNs have the same basic structure: they have a QSO sitting in the middle of a galaxy. Depending on whether we are looking down the barrel of a super-energetic jet, or right into the side of the gas torus, or at some angle in between, the QSO will have a different spectroscopic signature from our point of view. Furthermore, the QSO host galaxy could be spiral, elliptical, or peculiar, and we could be seeing the QSO through a screen of interstellar dust, or lots of gas, or a lot of stars of differing colors and luminosities. The host galaxy would therefore contribute its own components to the spectroscopic signature of an AGN. Depending on what is in the way, and what part of the QSO we can see, each AGN looks unique. Actually, they are all basically the same.

# MORE ACTIVE GALAXIES AND QUASARS

### How **many AGNs and QSOs** are there in the universe?

According to current observations, about 5 to 10 percent of all large galaxies in the nearby universe contain AGNs or QSOs. The brighter the QSO, the rarer it is. Only a small fraction of QSOs, for example, are as luminous as 3C 273. The farther back one goes in the history of the universe, however, the higher the incidence of QSOs becomes. This is an important piece of evidence that the universe is aging and evolving over time.

### If they are so uncommon, why are **AGNs and QSOs important** in the universe?

First of all, AGNs and QSOs are very energetic, often hundreds or thousands of times more luminous than any other galaxies in the universe. That means that they have a substantial influence on what goes on in their vicinity of the cosmos. QSOs, for example, may have played an extremely important part in the history of the universe some 12 billion years ago by ionizing (and thus rendering transparent) much of the obscuring interstellar gas spread throughout the cosmos at that time. Without this crucial ionization process, we would not be able to see through the foggy gas today, and astronomy would be a much more difficult occupation to pursue.

Secondly, current observations show that the vast majority of large galaxies in the universe contain supermassive black holes. That means that most galaxies have the raw ingredients to host an AGN or QSO, and possibly every large galaxy has undergone (or will undergo) AGN or QSO activity at some point in its lifetime. This makes them an extremely important part of the aging process of galaxies, so the more we understand them, the more we understand how the universe ages.

### How **bright** can such a **QSO searchlight** appear in the sky?

Here on Earth, no QSO or AGN is visible to the unaided eye. The brightest QSO as seen from Earth is 3C 273. It is about two billion light-years from Earth, which makes it a challenge for most small amateur telescopes to find. Compared to other distant objects, however, QSOs are brilliantly bright and relatively easy to detect with large astronomical telescopes. Several quasars known to be more than 11 billion light-years from Earth are more easily visible than the Sun would be if it were only 1,000 light-years away.

### What is a **quasar absorption line**?

If the spectrum of a quasar (or, more generally, an AGN or QSO) contains an absorption feature that was not produced by the quasar itself, that means that the quasar's light has shined through some material or object that absorbed some of that light. This kind of "quasar absorption line" can be studied by the effect it has on the quasar light, even if the absorbing object cannot be directly viewed by its own light emissions.

### What **causes** a **quasar absorption line**?

A quasar absorption line is usually caused by the interstellar medium within or surrounding a galaxy. The quasar's light goes through the medium, and the atoms in the medium absorb the quasar's light at specific wavelengths.

Occasionally, the interstellar medium that causes a quasar absorption line is associated with not a single galaxy, but a group or cluster of galaxies. It is also sometimes possible that the body of interstellar medium involved may be a large, free-

An artist's concept of two active galaxies with active nuclei containing black holes. The idea that galaxies without a central bulge like the one on the right could not contain black holes has been proven to be erroneous. (*NASA/JPL-Caltech*)

floating intergalactic cloud. One particular kind of quasar absorber is called a Lyman-alpha cloud, which is a cloud of intergalactic gas that is much smaller than a typical galaxy and has almost no dust or heavy elements in it.

## What is the **Lyman-alpha forest**?

When the spectrum of a QSO contains a very large number of absorption lines, the majority of those absorption lines are usually caused by Lyman-alpha clouds. These sub-galaxy-sized clumps of gas populate the distant universe at different redshifts. Each cloud produces a single absorption line caused by hydrogen atoms called the Lyman-alpha line (hence the name "Lyman-alpha cloud"). If there are enough Lyman-alpha clouds between us and the QSO, a large swath of the QSO spectrum can literally look "chopped up" by all the Lyman-alpha lines produced by these clouds, appearing at all different redshifts. The effect is that of a forest of trees sprouting up and down in the spectrum—hence the name Lyman-alpha forest.

## What can astronomers **learn** from the **Lyman-alpha forest**?

Since each absorption line in the Lyman-alpha forest of a QSO spectrum represents a single gas cloud, it is possible to count the number of Lyman-alpha clouds at each

87

redshift along the line of sight between the QSO and Earth. These clouds are not luminous enough to be observed directly, but they are important constituents of matter in the universe. Understanding the population of Lyman-alpha clouds, therefore, helps astronomers understand how gaseous matter is distributed throughout the cosmos. From studying the Lyman-alpha forest in numerous QSO spectra, astronomers have already deduced that there is about as much gaseous matter in the universe as there is stellar matter. In other words, the wispy and almost insubstantial interstellar and intergalactic medium is as significant a part of the cosmos as all the stars in all the galaxies in the universe.

# STARS

## STAR BASICS

### What is a **star**?

A star is a mass of incandescent gas that produces energy at its core by nuclear fusion. Most of the visible light in the universe is produced by stars. The Sun is a star.

### How **many stars** are there in the sky?

Without the interference of light from ground sources, a person with good eyesight can see about 2,000 stars on any given night. If both hemispheres are included, then about 4,000 stars are visible. With the help of binoculars or telescopes, however, the number of visible stars increases dramatically. In our Milky Way galaxy alone, there are more than 100,000,000,000 stars, and in our observable universe there are at least a billion times that number.

### What is the **closest star** to Earth?

The Sun is the closest star to Earth. It is 93 million miles away from Earth on average

### Other than the Sun, what is the **next closest star** to Earth?

The closest star system to Earth is the multiple star system Alpha Centauri. The faintest star in that system, known as Proxima Centauri, has been measured to be 4.3 light-years away from Earth. The main star in Alpha Centauri is about 4.4 light-years away. The table below lists other nearby stars.

## Stars Closest to the Sun

| Name | Spectral Type | Distance in Light-Years |
|------|---------------|-------------------------|
| Proxima Centauri* | M5V (red dwarf) | 4.24 |
| Alpha Centauri A* | G2V (sun-like) | 4.37 |
| Alpha Centauri B* | K0V (orange dwarf) | 4.37 |
| Barnard's Star | M4V (red dwarf) | 5.96 |
| Wolf 359 | M6V (red dwarf) | 7.78 |
| Lalande 21185 | M2V (red dwarf) | 8.29 |
| Sirius | A1V (blue dwarf) | 8.58 |
| Sirius B | DA2 (white dwarf) | 8.58 |
| Luyten 726-8A | M5V (red dwarf) | 8.73 |
| Luyten 726-8B | M6V (red dwarf) | 8.73 |
| Ross 154 | M3V (red dwarf) | 9.68 |
| Ross 248 | M5V (red dwarf) | 10.32 |
| Epsilon Eridani | K2V (orange dwarf) | 10.52 |
| Lacaille 9352 | M1V (red dwarf) | 10.74 |
| Ross 128 | M4V (red dwarf) | 10.92 |
| EZ Aquarii | M5V (red dwarf) | 11.27 |
| Procyon A | F5V (blue-green dwarf) | 11.40 |
| Procyon B | DA (white dwarf) | 11.40 |
| 61 Cygni A | K5V (orange dwarf) | 11.40 |
| 61 Cygni B | K7V (orange dwarf) | 11.40 |
| Struve 2398 A | M3V (red dwarf) | 11.53 |
| Struve 2398 B | M4V (red dwarf) | 11.53 |
| Groombridge 34 A | M1V (red dwarf) | 11.62 |
| Groombridge 34 B | M3V (red dwarf) | 11.62 |

*These stars are in the Alpha Centauri system.

## What is an **asterism**?

An asterism is a group of stars in the sky that, when viewed from Earth, create an outline of some recognizable shape or pattern. Two well-known asterisms are the Big Dipper, which many astronomers use to point out the location of the North Star, and the Summer Triangle, which is marked by three of the most prominent stars in the Northern Hemisphere's summer night sky.

# MAPPING THE STARS

## What is a **constellation**?

A constellation is akin to an asterism, but it is usually much more complicated, containing more stars or larger areas of the sky. A few asterisms are constellations: the asterism called the Southern Cross, for example, is the constellation Crux (the Cross). Modern constellations are mostly named after mythological themes, such as gods, legendary heroes, creatures, or structures. Although most constellations resemble the figures after which they are named, others are not as recognizable.

The constellations encompass the entire celestial sphere and provide a visual reference frame. Astronomers can plot the stars and other objects in the universe using constellations, charting the apparent movement that is caused by Earth's own rotation and orbit.

## How **many constellations** are there?

The current, internationally agreed-upon map of the sky contains 88 constellations. Some well-known constellations include Aquila (the Eagle), Cygnus (the Swan), Lyra (the Harp), Hercules and Perseus (two mythological heroes), Orion the Hunter and Ophiucus the Knowledge-seeker (two other mythological characters), Ursa Major and Ursa Minor (the Big Bear and Little Bear), and the constellations of the zodiac. The table below lists well-known constellations.

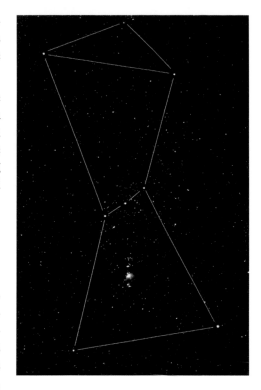

One of the most recognizable constellations is Orion the Hunter. (*Courtesy of Howard McCallon*)

### Well-Known Constellations

| Name | Common Name | Well-Known Stars in the Constellation |
| --- | --- | --- |
| Aquila | The Eagle | Altair |
| Auriga | The Charioteer | Capella |
| Bootes | The Hunter | Arcturus |
| Canis Major | The Big Dog | Sirius |
| Canis Minor | The Little Dog | Procyon |
| Carina | The Keel | Canopus |
| Crux | The Southern Cross | Acrux |
| Cygnus | The Swan | Deneb |
| Gemini | The Twins | Castor, Pollux |
| Leo | The Lion | Regulus |
| Lyra | The Harp | Vega |
| Orion | The Hunter | Rigel, Betelgeuse, Bellatrix |
| Ursa Major | The Big Bear | Dubhe, Alcor, Mizar |
| Ursa Minor | The Little Bear | Polaris (The North Star) |

The naming of constellations dates back to ancient civilizations. In 140 C.E. the ancient Greek astronomer Claudius Ptolemy cataloged 48 constellations visible from Alexandria, Egypt. All but one of those 48 are still included in present-day catalogs, and that one (Argo Navis, the Argonauts' Ship) was subdivided in the 1750s into four separate constellations. Many new constellations were named in later centuries, mostly in previously uncharted parts of the sky in the Southern Hemisphere. (Some of those constellations have since been abandoned.) Many of the constellations originally had Greek names; these names were later replaced by their Latin equivalents by which they are still known today.

## Who made the **first astronomical star catalogs** and **charts**?

Hipparchus, an ancient Greek astronomer of the second century B.C.E., is best remembered for his astronomical measurements and the instruments he created to make them. Hipparchus constructed an atlas of the stars visible without a telescope and categorized them by brightness. The first major astrometric satellite, which used parallax to measure the positions of and distances to more than 100,000 stars, was named *Hipparcos* in his honor.

Star catalogs increased dramatically in size after telescopes were invented. James Bradley (1693–1762) was England's Astronomer Royal from 1742 until his death 20 years later. He prepared an accurate chart of the positions of over 60,000 stars. German astronomer Johann Elert Bode (1747–1826), who became director of the Berlin Observatory in 1786, published an enormous catalog of stars and their positions in 1801.

## Who made the first **scientific map** of the **southern constellations**?

In 1676 English astronomer Edmund Halley (1656–1742) traveled to Saint Helena, an island off the west coast of Africa, and established the first European observatory in the Southern Hemisphere. There, he made the first scientific map of the southern constellations, recording the positions of 381 stars.

## What is the **astronomical significance** of **constellations**?

Scientifically, a constellation does not have any significance. Stars, nebulae, or galaxies in the same constellation may or may not have anything in common, aside from the fact that they are nearby in the sky as viewed from Earth. They may even be separated by a greater distance than objects in two different constellations.

That said, astronomers very often refer to objects being "in" or "toward" a certain constellation. That means—and *only* means—that those objects can be found by looking toward that particular constellation, as viewed from Earth. To say that a particular object is located "in" a constellation does not take into account at all the actual distance of that object from Earth, or from any other object in that constellation.

## What is the **North Star**?

The North Star is any star near the spot in the sky called the north celestial pole: the place that Earth's rotational axis is pointing toward. Right now, and for the past several centuries, a Cepheid variable star called Polaris has been very close to the pole, and thus has served as a good north star. Earth's rotational axis changes its pointing location across the sky over the millennia, however. Thousands of years ago, while ancient Egyptian culture thrived, the North Star was a dimmer star called Thuban. Between then and now, there have been stretches of many centuries when there was no useful North Star at all.

## Is there a **South Star**?

Right now, there is no easily visible star near the south celestial pole. There are many asterisms and celestial objects relatively near the pole, so it is possible to triangulate between them and roughly find the location of the south celestial pole.

# DESCRIBING AND MEASURING STARS

## What are the **brightest stars** in the night sky?

The brightest stars in the night sky as viewed from Earth are Sirius, the "Dog Star," in the constellation Canis Major (the Big Dog); Canopus, in the constellation Carina (the Keel); and Rigel Kentaurus, more commonly known as Alpha Centari, in the constellation Centaurus (the Centaur). These three stars are not, however, the three stars in the night sky that emit the most light; they are the three stars that emit the most light that reaches Earth.

The table below lists the brightest stars we can see from Earth.

### The Brightest Stars as Seen from Earth

| Name | Constellation | Spectral Type | Apparent V Magnitude | Distance (in light-years) |
|---|---|---|---|---|
| Sun | N/A | G2V (yellow dwarf) | −26.72 | 0.0000158 |
| Sirius | Canis Major | A1V (blue dwarf) | −1.46 | 8.6 |
| Canopus | Carina | A9II (blue giant) | −0.72 | 310 |
| Arcturus | Bootes | K1III (red giant) | −0.04(variable) | 37 |
| Alpha Centauri A | Centaurus | G2V (yellow dwarf) | −0.01 | 4.3 |
| Vega | Lyra | A0V (blue dwarf) | 0.03 | 25 |
| Rigel | Orion | B8I (blue supergiant) | 0.12 | 800 |
| Procyon | Canis Minor | F5V (blue-green dwarf) | 0.34 | 11.4 |

| Name | Constellation | Spectral Type | Apparent V Magnitude | Distance (in light-years) |
|------|--------------|---------------|---------------------|---------------------------|
| Achernar | Eridanus | B3V (blue dwarf) | 0.50 | 140 |
| Betelgeuse | Orion | M2I (red supergiant) | 0.58(variable) | 430 |
| Agena | Centaurus | B1III (blue giant) | 0.60(variable) | 530 |
| Capella A | Auriga | G6III (yellow giant) | 0.71 | 42 |
| Altair | Aquila | A7V (blue dwarf) | 0.77 | 17 |
| Aldebaran | Taurus | K5III (red giant) | 0.85(variable) | 65 |
| Capella B | Auriga | G2III (yellow giant) | 0.96 | 42 |
| Spica | Virgo | B1V (blue giant) | 1.04(variable) | 260 |
| Antares | Scorpio | M1I (red supergiant) | 1.09(variable) | 600 |

## How far are the **farthest stars**?

Among the 4,000 or so stars in the night sky that are visible to the unaided eye, the most distant among them are several thousand light-years away. The light from more distant stars is visible, however, when many of them are associated together in a star cluster or nearby galaxy. It is possible to see, for example, the combined starlight of the Large Magellanic Cloud (about 170,000 light-years away), the Small Magellanic Cloud (about 240,000 light-years away), or even the Andromeda galaxy (about 2.2 million light-years away) with unaided eyes. Using telescopes, we can see starlight from galaxies that are more than 12 billion light-years away.

## Who first **accurately measured** the **distance** to a **star**?

German mathematician and astronomer Friedrich Wilhelm Bessel (1784–1846) was 20 years old when he recalculated the orbit of Halley's comet and mailed his findings to astronomer Heinrich Olbers (the man famous for the paradox named in his honor). When Bessel was 26, he was appointed director of the Koenigsburg Observatory, a position which he held until his death in 1846. During his career, Bessel cataloged the positions of more than 50,000 stars. To study perturbations (small disturbances) of planetary motions in the solar system, he developed a series of mathematical equations that helped describe complex overlapping motions and vibrations. Today these equations are called the Bessel functions in his honor, and are indispensable tools in the fields of applied mathematics, physics, and engineering. Using innovative techniques, he measured the apparent motions of a large number of stars more accurately than ever before.

## How do astronomers **describe** the **brightness** of **stars**?

It is useful to describe the brightness of stars in terms of their flux, the measure of how much light arrives here on Earth from that star, or by their luminosity, the measure of how much energy they radiate. Astronomers, however, also use a historical description of a star's brightness known as magnitude.

Ancient Greek astronomers established the original magnitude system, whereby the brightest stars visible to the naked eye were categorized as "1st magnitude," the next brightest stars "2nd magnitude," and so forth. The faintest, barely visible stars were labeled "6th magnitude." After telescopes were invented, many more stars fainter than 6th magnitude were discovered. Astronomers thus extended the magnitude beyond first and sixth magnitudes, following a mathematical formula on a logarithmic scale.

Due to this historical origin of the magnitude system, brighter objects have a lower magnitude number, while fainter objects have a higher magnitude number. This means that negative magnitudes are brighter than positive magnitudes. Like a lot of things with long histories, the astronomical magnitude system is backwards and counterintuitive, but it works and persists to this day.

## What is the difference between **absolute magnitude** and **apparent magnitude**?

The original magnitude system is a flux-based system: the more light that reaches an observer on Earth, the lower its magnitude number. This is called apparent magnitude, because it is the apparent brightness of the star as seen from Earth.

The absolute magnitude system is a luminosity-based system: the more light that is emitted by a star regardless of where it is, the lower its magnitude number. It is defined as follows: the absolute magnitude of a star is what the apparent magnitude of the star would be if it were at a distance of 10 parsecs (about 32.6 light-years). Since flux and luminosity are related to one another by the distance between a light source and the observer, the difference between the apparent magnitude ($m$) and absolute magnitude ($M$) of a star is called its distance modulus ($m$–$M$).

# HOW STARS WORK

## Why do **stars shine**?

Stars shine because nuclear fusion occurs in their core. Nuclear fusion changes lighter elements into heavier ones and can release tremendous amounts of energy

## If nuclear fusion did not occur in the Sun, could it still shine?

For a while, the Sun could still shine without fusion. The Sun was originally formed when a large amount of matter fell toward a common center of gravity. As that matter compressed into a dense ball of gas, it grew very hot and began radiating heat and light—that is, it began to shine—even before nuclear fusion began to occur. If there were no nuclear fusion in the Sun, the collapse and compression of the Sun's gases would continue to generate energy until it all fell together into a single point.

According to calculations first made by Lord William Thomson Kelvin (1824–1907) and Hermann von Helmholtz (1821–1894) in the late-nineteenth century, this kind of energy generation by collapsing gases would have allowed the Sun to shine at its current luminosity for millions of years. But the energy could not have lasted the 4.6 billion years that we know the Sun has been shining. Without nuclear fusion, the solar system would have gone dark long before life first appeared on Earth.

in the process. The most powerful nuclear weapons on Earth are powered by nuclear fusion, but they are puny compared to the nuclear explosiveness of the Sun.

### How does **nuclear fusion work** in stars?

Atomic nuclei cannot just combine randomly. Rather, only a small number of specific fusion reactions can occur, and even then only under very extreme circumstances. In the Sun's core temperatures exceed 27 million degrees Fahrenheit (15 million degrees Celsius) and pressures exceed 100 billion Earth atmospheres. In these circumstances, there is a minute chance—less than a billion to one!—that, in any given year, a proton will fuse with another nearby proton to form a deuteron, also known as a deuterium nucleus or "heavy hydrogen" nucleus. The deuteron then fuses quickly with another proton to produce a helium-3 nucleus. Finally, after waiting around on average for another million years or so, two nearby helium-3 nuclei can fuse to form a helium-4 nucleus and release two protons.

In this multi-step sequence, called the "proton-proton chain," hydrogen is transformed into helium-4, and a tiny bit of matter is converted into energy. Even though it is very hard for any given pair of protons to fuse into a deuteron, there are so many protons in the core of the Sun that more than one trillion trillion trillion such fusion reactions occur there each second. The amount of mass converted into energy is thus huge—about 4.5 million tons per second—and provides enough outward push to keep the Sun in a stable size and shape, and shine its stellar glow out into space.

### Who first **explained** how **nuclear fusion works**?

Hans Albrecht Bethe (1906–2005) first explained the process of nuclear fusion. Born in Strasbourg, Germany, he studied in Britain and the United States, then

## Are stars solid, liquid, or gas?

**S**tars are mostly comprised of a special state of gas called plasma: gas that is electrically charged. Many people refer to plasma as the "fourth" state of matter. Other examples of plasma that we might observe in daily life include the air where a lightning bolt is traveling through, or the gas inside a fluorescent light bulb.

joined the physics faculty of Cornell University in 1935. There he worked on the theory of how quantum mechanical systems operate at high temperatures. In May 1938, Bethe published his findings explaining how nuclear fusion could work at the heart of the Sun, and how it could produce enough energy to make the Sun shine. Bethe's work in theoretical nuclear physics made him particularly valuable in the development of the first atomic bomb. He was deeply involved in the Manhattan Project during World War II, and was one of the pioneering scientists to work at Los Alamos National Laboratories in New Mexico. After the war, he continued to conduct pioneering research in the physics of stars and the processes that go on inside them. For his immense contributions to science, Hans Bethe was awarded the Nobel Prize in physics in 1967.

## Do **stars** have **electric currents** running through them?

Yes. The electric currents are far stronger than anything man-made, and they create the Sun's magnetic fields. There are some magnetic fields inside the Sun, and one very big field outside the Sun that extends into space for billions of miles.

## What happens in the **radiative zone** of the Sun?

The energy produced by nuclear fusion at the core of the Sun travels outward as radiation—photons traveling through the solar plasma. Although the photons travel at the speed of light, the plasma in a star is so dense that the photons keep running into particles and bouncing away in an unpredictable pattern called a random walk. The bouncing around is so extreme, that it takes an average of one million years for the solar light to travel the 250,000 miles (400,000 kilometers) through the radiative zone. In the vacuum of space, light can travel that distance in less than two seconds.

## What happens in the **convective zone** of the Sun?

The convective zone begins at a depth of about 90,000 miles (150,000 kilometers) below the surface of the Sun. In the convective zone, the temperatures are cool enough—under 1,800,000 degrees Fahrenheit (1,000,000 degrees Kelvin)—that the atoms in the plasma there can absorb the photons coming outward from the Sun's radiative zone. The plasma gets very hot, and begins to rise upward out of the Sun. The motion of the plasma creates convection currents, like those that happen in Earth's atmosphere and oceans, which carry the Sun's energy to the photosphere on seething rivers of hot gases.

This is how the convection works. As the temperature of the gas that has absorbed energy at the bottom of the convection zone increases, the gas expands, becoming less dense than its surroundings. These bundles of hot gas, because they are less dense, float up toward the surface of the convection layer like hot air balloons rising up into the air on a cold morning. At the top of this layer, they radiate away their excess energy, becoming cooler and denser, and then they sink down again through the convection layer. The effect is a continuous cycle of "conveyor belts" of hot gas moving up and cooler gas moving down.

## What happens in a star's **photosphere**?

The photosphere is the layer of a star's atmosphere that we see when viewing the Sun in visible light. It is sometimes referred to as the "surface" of a star. It is a few hundred miles thick, and it is made up of planet-sized cells of hot gas called granules. These gas cells are in constant motion, continuously changing size and shape as they carry heat and light through from the Sun's interior to its exterior. Sunspots, regions of intense magnetic activity, also occasionally appear in the photosphere and last from hours to weeks.

## What happens in a star's **chromosphere**?

The chromosphere, the thin and usually transparent layer of the Sun's atmosphere between the photosphere and the corona, is a highly energetic plasma that is punctuated with flares—bright, hot jets of gas—and faculae consisting of bright hydrogen clouds called plages. The chromosphere is generally not visible except with ultraviolet or X-ray telescopes.

The chromosphere is around 1,000 to 2,000 miles thick. It has some unexpected physical properties. For example, while the density of the gas decreases from the inner edge of the chromosphere to the outer edge, the temperature of the gas increases dramatically—from about 7,250 to 180,000 degrees Fahrenheit (4,000 to 100,000 degrees Celsius)—even though the distance to the Sun is actually increasing. At its outer limit, the chromosphere breaks up into narrow gas jets called spicules and merges into the Sun's corona.

## What **happens** in the Sun's **corona**?

The corona is a very thin, but very large, layer of gas that extends from a star's photosphere and chromosphere out to a distance of about 10 million miles away from the Sun. It is much dimmer than the rest of the Sun, and can only be seen when the Sun is blocked from view—either by a scientific instrument called a coronagraph, or naturally during a solar eclipse.

Even though it is thinner than the best laboratory vacuums on Earth and so far away from the Sun's core, the corona is very energetic and very hot, with its plasma reaching temperatures of millions of degrees. Astronomers are still trying to figure out how the corona gets so hot. Current research suggests that the strong electrical currents and magnetic fields in and around the Sun transfer tremendous amounts of energy to the corona, either generally or by special "hotspots" that form for short periods of time and then disappear again.

### Do **other stars** have layers like a **core, radiative zone, convective zone, photosphere, chromosphere,** and **corona**?

Yes, but in different ratios of thickness depending on the star's temperature, mass, and age. Very hot, young stars can even be completely radiative and have no convective zone; very cool stars, on the other hand, can be completely convective and have no radiative zone. The coronae around stars can also vary tremendously, depending on the strengths of the magnetic fields around the stars.

# SUNSPOTS, FLARES, AND SOLAR WIND

### What is a **sunspot**?

Sunspots, when viewed by visible light, appear as dark blemishes on the Sun. Most sunspots have two physical components: the umbra, which is a smaller, dark, featureless core, and the penumbra, which is a large, lighter surrounding region. Within the penumbra are delicate-looking filaments that extend outward like spokes on a bicycle wheel. Sunspots vary in size and tend to be clustered in groups; many of them far exceed the size of our planet and could easily swallow Earth whole.

Sunspots are the sites of incredibly powerful, magnetically driven phenomena. Even though they look calm and quiet in visible light, pictures of sunspots taken in ultraviolet light and in X rays clearly show the tremendous energy they produce and release, as well as the powerful magnetic fields that permeate and surround them.

## What is a **solar prominence**?

Prominences are high-density streams of solar gas projecting outward from the Sun's surface (photosphere) into the inner part of the corona. They can be more than 100,000 miles long and can maintain their shapes for days, weeks, or even months before breaking down.

## What is a **solar flare**?

Solar flares are sudden, powerful explosions on the surface of the Sun. They usually occur when large, powerful sunspots have their magnetic fields too tightly twisted and torqued by the hot, swirling plasma in the Sun. The magnetic field lines unwind and break suddenly, and the matter and energy that had been contained rushes outward from the Sun. Solar flares can be many thousands of miles long, and they can contain far more energy than all of the energy consumption of all of human history on Earth.

## What is a **coronal mass ejection**?

A coronal mass ejection is a huge blob of solar material—usually highly energetic plasma—that is thrown outward into space in a huge solar surface explosion. Coronal mass ejections are associated with solar flares, but the two phenomena do not always occur together. When coronal mass ejections reach the space near Earth, artificial satellites can be damaged by the sudden electromagnetic surge caused by the flux of these charged particles.

## What is the **solar wind**?

The solar wind is the flow of electrically charged particles outward from the Sun. Aside from stormy outbursts like solar flares, it streams gently from the Sun's corona throughout the solar system. The solar wind can vary in its speed and intensity, just like wind on Earth; it is, however, streaming plasma and not moving air.

## How do we see the **effects** of the **solar wind** in the **solar system**?

One easily visible effect of the solar wind can be seen in the tail of a comet. When a comet enters the inner solar system, the increased temperatures cause it to lose a

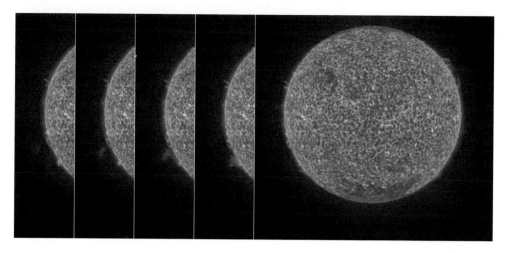

An 80,000-mile-long solar flare erupts from the Sun in this image taken by NASA's Solar and Heliospheric Observatory. (*NASA*)

small portion of its outer layers, which sublimates from solid to gas. The loosened material is swept back away from the Sun, forming the comet's tail. The electrically neutral particles are pushed back by the Sun's radiation pressure—the momentum of sunlight itself—while the electrically charged particles are pushed back by the solar wind. Sometimes, these two components separate slightly, and we can see both a "dust tail" and an "ion tail."

## How **fast** does the **solar wind** travel?

The flow of plasma out from the Sun is generally continuous in all directions, typically moving at speeds of several hundred kilometers per second. It can, however, gust out of holes in the solar corona at 2,200,000 miles per hour (1,000 kilometers per second) or faster. As the solar wind travels farther from the Sun, it picks up speed, but it also rapidly loses density.

## How **far** does the **solar wind** travel?

The Sun's corona extends millions of miles beyond the Sun's surface. The plasma of the solar wind, however, extends billions of miles farther—well beyond the orbit of Pluto. Beyond there, the plasma density continues to drop. There is a limit, called the heliopause, where the influence of the solar wind dwindles to just about nothing. The region inside the heliopause—which is thought to be some 8 to 14 billion miles (13 to 22 billion kilometers) from the Sun—is called the heliosphere.

## Do **all stars** have **spots, prominences, flares, mass ejections,** and **winds**?

Yes, in varying degrees all stars have these characteristics. The Sun, compared to most stars we know, is relatively quiet in its stormy activity. That is very good news to living things on Earth, which generally cannot survive too much disruption. Some stars have huge flares constantly erupting from them; others actually have most of their surface covered with spots. If there were planets orbiting those stars, the electromagnetic results in the environments there would almost certainly not support life as we know it.

# STAR EVOLUTION

### What is **stellar evolution**?

Stellar evolution is the term used to describe the aging process of stars. The theory of stellar evolution is broad and complicated, and is one of the most important ideas in all of astronomy. It is remarkably analogous to the study of aging in humans: we are born, go through immature stages, then are mature for a long time, and then undergo further and final changes toward the end of our lives until we finally die.

### What is a **main sequence star**?

A "main sequence star" is a star that is currently in the main mature period of its life cycle. Main sequence stars are converting hydrogen into helium and are in an equilibrium state.

### Are there **stars not** on the **main sequence**?

Yes. Whereas most stars in any given population of stars are on the main sequence—that is, going through the longest equilibrium period of their life cycle—a small percentage of stars in that population are not. This includes pre-main sequence, or "infant" stars, and post-main sequence, or "elderly" stars. Stars change and age throughout their existence.

## How do astronomers use the **H-R diagram** to **study** populations of **stars**?

The H-R (Hertzsprung-Russell) diagram plots the luminosity or magnitude of stars on the vertical axis and the photospheric temperature, color, or spectral type of those same stars on the horizontal axis. In a typical population of stars, the vast majority of stars will appear on a narrow diagonal band called the main sequence; the sequence runs from hot-and-luminous stars to cool-and-dim stars. Stars that are cool yet luminous, usually red giant stars, are not on the main sequence; stars that are hot yet dim, usually white dwarf stars, are also not on the main sequence. Stars not on the main sequence are generally in the end of their life cycles, and their locations on the diagram indicate the stage of life cycle they are in.

Among the many ways to analyze the H-R diagram, looking at the bright and dim limits of the main sequence can help determine the age of the population; the number of different kinds of non-main sequence stars can help determine the evolutionary history of the population; and an extra band of stars parallel to the main sequence could indicate the presence of a second population of star mixed in with the first. Almost every detail of where the data points are on the H-R diagram can provide a valuable piece of evidence about the nature of a complex stellar population.

## What is a **color-magnitude diagram**?

A color-magnitude diagram is a type of Hertzsprung-Russell diagram that plots the apparent magnitude of stars on the vertical axis and the color of those same stars on the horizontal axis. They are particularly useful for studying populations of stars in clusters.

## What is a **Wolf-Rayet star**?

Named after the two astronomers who discovered the first example of this class of object, a Wolf-Rayet star is a high-mass star that is very young. It is pretty much a main sequence star, but it is so young that it has not reached a steady equilibrium; very strong stellar winds are gusting off the surface of the star, creating a wildly fluctuating, dynamic environment.

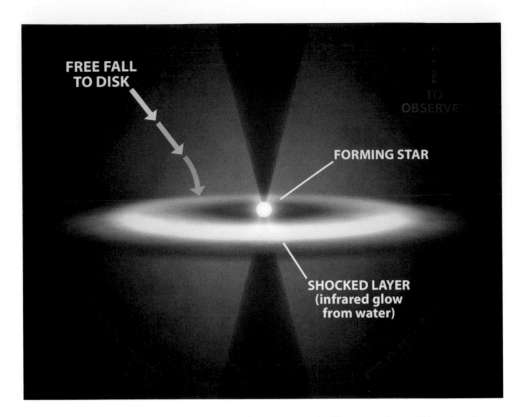

FREE FALL TO DISK

TO OBSERVER

FORMING STAR

SHOCKED LAYER
(infrared glow
from water)

A diagram illustrating the formation of a protoplanetary disk containing considerable amounts of water. Disks such as these are easier to detect from Earth when the disk is oriented face-on toward the observer. (*NASA/JPL-Caltech/T. Pyle*)

## What is a **T Tauri star**?

Named after the first object of its class, a T Tauri star is an intermediate-mass star that is very young. It is probably so young that nuclear fusion has not yet begun at its core, or maybe has just begun. The material surrounding the core has not yet settled into equilibrium, so much of it is still falling in toward the star's center. Meanwhile, the infall is creating huge amounts of energy that comes outward from the center in the form of strongly gusting stellar winds. Because of this the center of the star is obcured from view by all the swirling dust and gas.

## What is a **protostar**?

A protostar is the name given to the class of objects that are stars that are not quite on the main sequence. In other words, one might call them "baby stars." A T Tauri star can be considered an example of a protostar.

## What is a **protoplanetary disk**?

Once a star has begun sustained nuclear fusion at its core, its stellar winds start to clear out the surrounding dust, gas, and other debris. Some of that debris, however, settles into a thin, swirling disk that orbits around the newly born star. This

> ## What is the most important factor that influences how a star will evolve?
>
> **A** star's initial mass—the mass of a star when it is born—is by far the most important factor that influences its evolution (i.e., the aging process). Very generally speaking, stars fall into five mass categories: very low mass (down to about 0.01 solar mass), low mass (about 0.1 solar mass), intermediate mass (about 1 solar mass), high mass (about 10 solar masses), and very high mass (up to about 100 solar masses). Each of these categories follows a generally similar path from starbirth to star-death. The Sun, which by definition has one solar mass, is thus an intermediate-mass star.

structure is referred to as a protoplanetary disk. It is so named because that is where the raw materials for the planets in that star system have gathered, and where those planets themselves will likely be born.

## What is the **relationship** between the **initial mass** of a star and its **size, age,** and **luminosity**?

The main part of a star's life cycle is spent on what is called the main sequence. The higher the initial mass of a star is, the greater its main sequence luminosity is; the bluer and hotter it is; the larger its diameter is; and the shorter its main sequence lifetime is.

## How does a **very low-mass star evolve**?

A very low-mass star is often called a brown dwarf. It is born, lives, and ultimately dies in almost exactly the same form. A typical brown dwarf, containing one hundredth the mass of the Sun, has a luminosity about one-millionth that of the Sun. It will shine, albeit feebly, for a hundred trillion years or more.

## How does a **low-mass star evolve**?

A low-mass star is sometimes called a red dwarf. It is born fusing hydrogen into helium; it continues to do so until it stops, never really changing size and form during that time. It ends its life cycle as a white dwarf. A typical low-mass star, containing one-tenth the mass of the Sun, has a luminosity about one thousandth that of the Sun, and a main sequence lifetime of about one trillion years.

## How does an **intermediate-mass star evolve**?

A star about the mass of the sun, also called an intermediate-mass star, is born fusing hydrogen into helium. After it goes through its main sequence lifetime, it undergoes a dramatic change, becoming a red giant for a relatively short time.

105

## What is a supernova remnant?

**A** supernova remnant is the glowing emission nebula that is left over after a supernova explosion. It is comprised of the plasma that used to be part of the massive star which was blown apart. The remnant originally is pushed outward into space at a speed of up to 100 million miles per hour. Over time, the remnant forms bright filaments of highly energized gas. Furthermore, this gas is highly enriched with heavy elements, the result of the nuclear fusion right near the end of the progenitor star's life. These elements, such as calcium, iron, and even silver and gold, wind up being incorporated into the interstellar medium and become the raw materials for future generations of stars and planets. The Crab Nebula is a famous example of a supernova remnant.

Eventually, the star finishes the red giant phase and collapses into a white dwarf, its final configuration. The Sun, which contains one solar mass of material and emits one solar luminosity unit of light, will have a main sequence lifetime totaling about ten billion years. It will then be a red giant for about one-tenth that amount of time.

## How does a **high-mass star evolve**?

A high-mass star starts out its life as a luminous main sequence star, and also later becomes a red giant. Instead of collapsing and fading into a white dwarf, however, it fuses not only hydrogen into helium, but also helium into carbon, carbon into oxygen, and so forth. This creates heavier and heavier elements, including neon, magnesium, silicon, and iron. Then, when the equilibrium between the inward pull of gravity and the outward push of nuclear fusion energy is broken, the star's own gravity collapses the core of the star in a tiny fraction of a second, blowing itself apart in a titanic explosion called a supernova. The final remnant of this evolutionary path is a neutron star. A neutron star is the collapsed stellar core and is only about 10 miles across, yet several times more massive than the Sun. A high-mass star that contains about 10 times the Sun's mass, in fact, would be about one thousand times more luminous during its main sequence and would have a main sequence lifetime of about 100 million years.

## How does a **very high-mass star evolve**?

A very high-mass star fuses hydrogen into helium fast and furiously. Having 100 times the mass of the Sun, these stars have a main sequence lifetime of about one million years and a luminosity a million times that of the Sun. Like a high-mass star, a very high-mass star leaves the main sequence and fuses heavier and heavier elements. When the supernova explosion occurs, however, the core does not stop collapsing at a neutron star. Rather, the mass of its core is so great—up to 10 or 20 solar masses—that no kind of ordinary matter can arrest the gravitational infall. The mass piles into a singularity and becomes a black hole.

## What is a **planetary nebula**?

Planetary nebulae, though their name seems ambiguous, are really clouds of gas. They are called "planetary" because when they were first discovered astronomers saw these nebulae as round and colorful; they looked like the planets in our solar system. A planetary nebula is produced by an intemediate-mass, Sun-like star going through the final stages of its life cycle. As such a star evolves past the red giant stage, the outer gaseous layers detach from the stellar core in a series of violent "puffs," shedding the atmosphere. Some well-known planetary nebulae include the Ring, the Cat's Eye, the Hourglass, and the Helix.

Located in the Large Magellanic Cloud, Hodge 301 is a cluster of dying stars surrounded by the Tarantula Nebula. Many of the stars within Hodge 301 have either exploded as supernovas or are aging red giants that will soon explode. (*NASA, The Hubble Heritage Team, STScI, AURA*)

## What is a **supernova**?

A supernova is a tremendous explosion that occurs when the core of a star exceeds the Chandrasekhar limit, and its collapse is not halted by electron degeneracy. When that happens, it takes only a fraction of a second for the stellar core to collapse into a dense ball about ten miles across. The temperature and pressure becomes almost immeasurably hot and high; and the recoil of that collapse causes an enormous detonation. More energy is released in ten seconds than the Sun will emit in its entire ten billion year lifetime, as the guts of the star are blown outward into interstellar space.

There are two general types of supernovae. A Type I supernova is the result of an existing, older white dwarf that gains enough mass to exceed the Chandrasekhar limit, causing a runaway collapse. A Type II supernova is produced by a single high-mass star whose gravity is so strong that its own weight causes the stellar core to reach a mass beyond the Chandrasekhar limit.

# THE SUN

## How **bright** is the **Sun** compared to other stars?

The apparent magnitude of the Sun is a large negative number. As viewed in visible light, the Sun has $m = -26.7$ brightness because it is so close and, thus, has the lowest apparent magnitude of any celestial object. The Sun's absolute magnitude is 4.8 as viewed in visible light. This number, unlike the Sun's apparent magnitude, is roughly in the middle of the range of most stars.

The Sun is the closest star to Earth. About 93 million miles away, it is over 100 times larger than the Earth. (*NASA/JPL-Caltech/R. Hurt*)

## How **long** has the **Sun** been **shining**?

The Sun has been shining for 4.6 billion years. We know this from a variety of scientific studies. The most convincing evidence comes from the study of meteorites. Using various dating methods, some of these meteorites have been shown to have formed at the time the Sun began to shine. They have been dated to be 4.6 billion years old, so the Sun is estimated to be that old, as well.

## How **much longer** will the **Sun shine**?

Based on the scientific understanding of how stars work, our Sun will continue to conduct nuclear fusion at its core for about another five to six billion years.

## What is the **size** and **structure** of the **Sun**?

The Sun has a core at its center; a radiative zone surrounding the core; a convective zone surrounding the radiative zone; a thin photosphere at its surface; and a chromosphere and corona that extends beyond the photospheric surface. In all, the Sun is about 853,000 miles (1,372,500 kilometers) across, which is about 109 times the diameter of Earth.

The different zones and layers in and around the Sun exist because the physical conditions—mostly temperature and pressure—of the Sun change depending on the distance from the Sun's center. At the core, for example, temperatures exceed 15 million degrees Kelvin, whereas the inner part of the convective zone is just under 1 million degrees Kelvin, and the photosphere is about 5,800 degrees Kelvin.

## What is the **Sun made of**?

The Sun's mass is composed of 71 percent hydrogen, 27 percent helium, and 2 percent other elements. In terms of the number of atoms in the Sun, 91 percent are hydrogen atoms, 9 percent are helium atoms, and less than 0.1 percent are atoms of other elements. Most of the stars in the universe have a similar chemical composition.

## How **massive** is the **Sun**?

The Sun has a mass of 4.39 million trillion trillion pounds (1.99 million trillion trillion kilograms). The most massive supergiant stars have about one hundred times more mass than the Sun. The least massive dwarf stars and brown dwarfs contain about one-hundredth the mass of the Sun.

## How **hot** is the **Sun**?

The temperature at the center of the Sun is about 27 million degrees Fahrenheit (15 million degrees Kelvin). This is typical for stars that convert hydrogen into helium using the proton-proton chain, but it is hotter than some stars and much cooler than others. This is expecially true if these other stars harbor fusion processes other than the proton-proton chain, such as the carbon-nitrogen-oxygen cycle or the triple-alpha reaction.

The temperature at the surface of the Sun is about 11,000 degrees Fahrenheit (5,800 degrees Kelvin). The surface temperatures of stars range typically from about 5,400 to 54,000 degrees Fahrenheit (3,000 to 30,000 degrees Kelvin), though in some special kinds of stars the surface temperatures can be higher or lower than this range.

## Does the **Sun spin**?

The Sun does indeed spin, rotating about its axis from west to east, the same direction that the planets orbit around the Sun. Since the Sun is not a solid object but rather a big ball of electrically charged gas, it spins at different speeds depending on the latitude. The Sun spins once around its axis near its equator in about 25 days, and in about 35 days near its north and south poles. This kind of spinning, in which different parts move at different speeds, is called differential rotation.

## What are the **consequences** of the **Sun's spin**?

Magnetic fields in the Sun, created by strong electric currents, are produced because of the Sun's spin. The Sun has differential rotation, and its interior roils with tremendous heat and energy. That causes the magnetic field lines in the Sun to get bent, twisted, knotted, and even broken; sunspots, prominences, solar flares, and coronal mass ejections are the result.

## Do **other stars spin**?

All stars spin at least somewhat. Whereas the Sun takes several weeks to rotate once on its axis, some stars can make a full rotation every few days. Stellar remnants, such as white dwarfs and neutron stars, can rotate even faster—some neutron stars rotate hundreds of times per second.

# DWARF STARS AND GIANT STARS

## What is a **brown dwarf**?

A brown dwarf is another name for a very low-mass star. The existence of brown dwarfs—stars with so little mass that there is almost no nuclear fusion in them, yet with much more mass than any planet in our solar system—was not confirmed until the 1990s. The reason is that their photospheres are so cool that they are very dim, emit very little visible light, and can be found only using infrared telescope technology. Since their discovery, infrared telescopes and infrared astronomical cameras have advanced by leaps and bounds. One result is that a huge number of brown dwarfs have been discovered in recent years. In fact, so many have been identified that it is now hypothesized that the number of brown dwarfs may outnumber all the other stars in our galaxy put together.

In this artist's depiction, our solar system is compared to what a brown dwarf star system might look like. (*NASA/JPL-Caltech/ T. Pyle*)

## What was the first white dwarf ever detected?

In the early twentieth century, astronomers studying the star Sirius (the Dog Star and brightest star in the night sky as viewed from Earth) noticed a tiny companion near the bright star. This companion, Sirius B, orbited around Sirius at a very small distance. By measuring the tiny wobbles in their mutual orbit, they deduced that Sirius B was more massive than our Sun, but smaller than Earth. Sirius B is the first white dwarf ever detected, and it remains one of the most massive white dwarfs known to astronomers.

## What is a **red dwarf**?

A red dwarf is another name for a low-mass, main-sequence star. They are cool compared to most other kinds of stars (their photospheric temperature is about 6,000 degrees Fahrenheit, or 3,000 degrees Kelvin), so they glow a dull red. Red dwarfs are small and faint compared to most other kinds of stars.

## What is a **red giant**?

A red giant is a kind of star that represents an evolutionary phase of intermediate and high-mass stars that have surpassed their main sequence lifetimes. When a star like the Sun becomes a red giant, a sudden burst of energy is produced by new fusion processes at the core of the star. This burst pushes the plasma in the star outward. When the equilibrium of the star's inward and outward forces are restored, the star has swelled to about one hundred times its original diameter. The swollen, bloated star is so large that its outer layers do not contain as much star-stuff, and the star's surface (photosphere) cools down to the temperatures of red dwarfs (about 6,000 degrees Fahrenheit or 3,000 Kelvin). The Sun is destined to become a red giant, and when it does, about five billion years from now, it will swallow the planets Mercury and Venus, and destroy Earth as well.

## What is a **white dwarf**?

A white dwarf is one common kind of stellar "corpse." Stars of intermediate and low mass tend to end their lives as white dwarfs. As the energy produced by nuclear fusion dwindles and ends in the cores of these stars, they collapse under their own weight until the atomic nuclei in the stars' plasma bump up against one another. Any further collapse of the star is halted by the atoms pushing against one another: a condition called electron degeneracy. The collapse concentrates the remaining heat of the dying star into a tiny space, causing the white dwarf to glow white-hot. A white dwarf the mass of the Sun will only be as large as our planet Earth, a shrinkage of about 100 times in diameter and a million times in volume. One teaspoon of white dwarf star material weighs several tons.

In this artist depiction of the binary system 4U 0614+091, material from a white dwarf is sucked into the gravity well of a pulsar. (*NASA/JPL-Caltech/R. Hurt*)

### Who **first described** the nature of **white dwarfs**?

The British theoretician Arthur Stanley Eddington (1882–1944) was the most distinguished astrophysicist of his time. He was the first scientist to propose that the tremendous heat production at a star's core is what prevents a star from collapsing under its own gravity. His seminal book, *The Internal Constitution of the Stars,* helped launch the modern theoretical study of stellar evolution. When astronomers puzzled over the nature of Sirius B, Eddington suggested the explanation that turned out to be correct: the matter of Sirius B is in a state called electron degeneracy—a special condition that is not found anywhere on Earth.

### Who **first** suggested that **some stars** could **not end their lives** as **white dwarfs**?

The Indian-American astrophysicist Subramanyan Chandrasekhar (1910–1995) first proposed this idea. In 1936, Chandra was hired to teach at the University of Chicago and to conduct research at Yerkes Observatory in Wisconsin. Over a long and remarkable career in Chicago, he made major advances in theoretical astrophysics, including work on the transfer of energy in stars and throughout the universe. He served as editor-in-chief of the *Astrophysical Journal* for a generation, as well. Chandra is perhaps best known, however, for discovering that stars can evolve beyond white dwarfs to other, even denser states of matter. He is widely regarded as the leading astrophysicist of his time.

## What is the **Chandrasekhar limit**?

In 1930 Chandrasekhar used theories first presented by Arthur Eddington, as well as Albert Einstein's special theory of relativity, to calculate that a star higher than a certain mass limit will not end its life as a white dwarf. In other words, the electron degeneracy that would stop the collapse of a star's core would stop working because the pressure would be so great that the electrons would start moving too fast to provide outward pressure. In 1934 and 1935, he made further calculations showing that, above about 1.4 times the mass of the Sun, a stellar core will collapse beyond the white dwarf stage and turn into something far denser and more compact. Although this particular discovery was not immediately accepted by the astrophysical community, the discovery of the Crab Nebula pulsar and the realization that it was far smaller and denser than any white dwarf confirmed Chandrasekhar's calculations. That upper mass limit is today called the Chandrasekhar limit in his honor.

## What is a **blue giant**?

A blue giant is the name for a star that is, as its name suggests, big and blue. Such stars are usually high-mass stars on the main sequence. Blue giants live for only a million years or so, glowing a million times brighter than the Sun before they blow apart in titanic supernova explosions.

# NEUTRON STARS AND PULSARS

## What is a **neutron star**?

A neutron star is the collapsed core of a star that is left over after a supernova explosion. It is, so to speak, matter's last line of defense against gravity. In order to stay internally supported as an object and not be crushed into a singularity, the neutrons in the object press up against one another in a state known as neutron degeneracy. This state, which resembles the conditions within an atomic nucleus, is the densest known form of matter in the universe.

## How **dense** is a **neutron star**?

A neutron star is about as dense as a neutron itself. To put it in a different way, it has the density of an object more massive than the Sun, yet it is only about ten miles across. That means that a neutron star is 10 trillion times denser than water. A single teaspoon of neutron star material would weigh about five billion tons! A dime-sized sliver of neutron star material contains more mass than every man, woman, and child on Earth put together. If one dropped a chunk of neutron star material toward the ground, it would cut through our planet like it was not there;

it would fall through the center of our planet, emerge out the other side, and keep traveling back and forth through the middle of Earth for billions of years, turning our planet into something like a big ball of Swiss cheese.

### What is the **environment** like around a **neutron star**?

The gravitational well of a neutron star is pretty steep. The effect on spacetime near the surface of a neutron star is therefore significant; objects in the sky would look distorted and displaced, and their colors would be gravitationally redshifted. If matter falls onto a neutron star, what happens is very similar to matter falling onto a black hole; the material does not disappear forever, but it certainly gets very hot, and can glow with X rays, ultraviolet radiation, and radio waves. If the neutron star is spinning as well, then a magnetic field billions of times stronger than Earth's can be created, causing highly energetic and radiative effects.

### What is a **pulsar**?

When a neutron star spins, it sometimes spins incredibly fast—up to hundreds of times a second. A magnetic field billions of times stronger than Earth's can form as a result. If the field interacts with nearby electrically charged matter, it can result in a great deal of energy being radiated into space, a process called synchrotron radiation. In this scenario, the slightest unevenness or surface feature on the neutron star can cause a significant "blip" or "pulse" in the radiation being emitted. Each time the neutron star spins around once, a pulse of radiation comes out. Such an object is called a pulsar.

### Who **first discovered** a **pulsar**?

In the 1960s, an astronomy graduate student at Cambridge University named Jocelyn Susan Bell Burnell (1943–) and her advisor Antony Hewish (1924–) used a large radio telescope in their research. The giant radio telescope consisted of scraggly looking antennae linked by wires, spread over a four-acre field, and was capable of detecting faint and rapidly changing energy signals and recording them on long rolls of paper. In 1967 Bell Burnell noticed some strange signals being recorded: periodic pulses of radio waves coming from specific locations in the sky. She found four pulsating sources; they were very mysterious because, prior to that time, the only recorded radio signals coming from space were continuous ones. Bell Burnell and Hewish hypothesized that these "pulsars" might be rapidly spinning white dwarf stars or neutron stars. The interpretation that they are neutron stars was eventually confirmed.

### How many pulsars have been discovered?

As of 2008, more than 1,000 pulsars have been found throughout our galaxy. Perhaps the best known one is the Crab Nebula pulsar. It is at the center of the Crab Nebula and is a remnant from a supernova that was first observed in 1054 C.E. It pulses once every 33 milliseconds; it is remarkable to imagine a body the mass of the Sun spinning more than 30 times per second!

# RADIATING STARS

### What is an "X-ray star"?

An "X-ray star," as its name implies, is a star that emits a great deal of X-ray radiation. Our Sun, as with most typical stars, emits lots of X rays compared to terrestrial sources. As a percentage of the total radiation emitted by the Sun, however, its X-ray emission is very small. X-ray stars may emit thousands of times more X rays than visible light radiation.

X-ray stars are almost always binary star systems or multiple star systems. The interaction between the two or more stars in the systems—one of which is usually a compact object like a white dwarf, neutron star, or black hole—is what causes the strong X-ray emission. Astronomers usually use the terms "low-mass X-ray binary" (LMXRB) or "high-mass X-ray binary" (HMXRB) to describe the two main classes of X-ray star systems.

### What's the difference between a low-mass X-ray binary and a high-mass X-ray binary?

As their names imply, a low-mass X-ray binary contains stars that are of relatively low mass, being of intermediate mass or lower, with a white dwarf as the compact companion. A high-mass X-ray binary, by contrast, often has one or two high-mass or very high-mass stars in the system, and the compact object is usually a neutron star or black hole. Though both systems emit copious amounts of X-ray radiation, their X-ray spectral signatures differ substantially because the physical conditions in those binary star systems are affected in different ways by the masses of the stars themselves.

**How did an X-ray binary lead to the discovery of the first confirmed stellar black hole?**

The most powerful X-ray source in the direction of the constellation Cygnus (The Swan) is called Cygnus X-1. After it was discovered, astronomers used various methods of observation to study this enigmatic object. It was discovered that Cygnus X-1 is a high-mass X-ray binary, but the compact object in the binary system was simply invisible. Furthermore, measurements of the motion of the other star in the binary—an impressive high-mass star in its own right—showed that the compact component was far more massive than any white dwarf or neutron star could possibly be without violating the laws of physics. In the end, the evidence was overwhelming that Cygnus X-1 contained a stellar black hole at least ten times the mass of our Sun.

## What was the **first X-ray binary star** ever **discovered**?

The first X rays from an astronomical source were detected by an X-ray telescope that was launched into space in 1962. The X rays seemed to come from the direction of the constellation Scorpius, but astronomers could not pinpoint exactly where in the constellation the emission came from. The source was given the name Scorpius X-1 (meaning the most powerful X-ray source in the direction of Scorpius). Over time, better technology and careful observations showed that the X rays were coming from an X-ray binary star system.

## What is a **polar**?

Not "polar" as in "polar bear," a polar (POE-larr) is the nickname for a kind of star with a high level of polarized light coming from it. In space, light becomes polarized when countless numbers of crystalline dust grains are aligned by strong magnetic fields to face a single direction. Together, they act like a huge cloud of microscopic mirrors and reflect polarized light in a specific proportion. By comparing the amount and orientation of polarized versus unpolarized light, it is possible to determine the configurations of the super-strong magnetic fields around stars that make such a phenomenon possible.

It turns out that polars are binary star systems, usually cataclysmic variables or even low-mass X-ray binaries. The magnetic fields that create the polar phenomenon are millions to billions of times the strength of the Sun's magnetic field and cause fascinating physical consequences in the binary system.

## What is a **gamma-ray burst**?

About once a day, a flash of gamma-ray radiation reaches Earth from far out in space. Some of these gamma-ray bursts occur within our own Milky Way galaxy; others occur in galaxies far, far away. Some gamma-ray bursts have been detected over 10 billion light-years away! Gamma rays are the most energetic type of electromagnetic radiation, and stars rarely emit large amounts of them.

Some gamma-ray bursts—especially those within our galaxy—appear to be caused by explosive detonations of some kind in binary star systems. Usually, one or both of the stars in these systems are dense, massive stellar end-products like white dwarfs, neutron stars, or black holes. The gamma-ray bursts observed in distant galaxies could be caused by the collision of neutron stars and/or black holes. Alternately, when a massive star explodes as a supernova just as it is spinning rapidly, the combination of stellar collapse and stellar rotation can emit two super-powerful, tightly focused beams of gamma rays outward into space. These beams are carrying more radiation than the Sun makes in millions, even billions, of years.

# BINARY STAR SYSTEMS

## What is a **binary star**?

A binary star is a pair of stars that are so close together in the sky that they appear to be closely associated with one another. Some binary stars, called apparent binaries, are merely close together because of our point of view from Earth; they have nothing to do with one another physically. When two stars that are physically associated together make a binary star system, however, the two stars orbit each other around a single center of gravity.

Physically associated binary stars are further divided into categories. A visual binary is a pair where each star can be observed distinctly, either through a telescope or with the unaided eye. An astrometric binary is a pair where the two stars cannot be distinguished visually, but the wobble of one star's orbit indicates the existence of another star in orbit around it. An eclipsing binary is a pair where the plane of the stars' orbit is nearly edgewise to our line of sight; the stars take turns being partially or totally hidden by one another. A spectroscopic binary is a pair where two stars can be detected by Doppler shifts or other spectral indicators from spectroscopic measurements.

There are also multiple star systems, which may have three or four stars orbiting one another around a single center of gravity, although they are rarer and less likely to be in a long-term stable orbit.

## Who made the **first catalogs** and **charts** of **binary stars**?

The German astronomer William Herschel (1738–1822), who lived and worked in England, mapped out 848 pairs of binary stars, showing that the force of gravity acts between stars, as theorized by Isaac Newton. He hypothesized that stars originally were randomly scattered throughout the universe, and that over time they came together in pairs and clusters.

Astronomers have learned that stable, mature planetary systems might be more common around binary stars, thus making a sunset such as this artist-illustrated one less exotic than we might think. (*NASA/JPL-Caltech/R. Hurt*)

### How **common** are **binary stars** and **multiple stars**?

In the part of the Milky Way galaxy where the Sun resides, at least half of the stars have been shown to be in binary and multiple star systems. The actual fraction of stars that are binary or multiple is not exactly known, and this is still a subject of frontier scientific research. Certainly, the fraction is high enough that it is an important factor to take into account when astronomers study stellar birth and life cycles.

### What is an **AM Herculis star**?

An AM Herculis star, named after the first example of this kind of object ever discovered, is a special kind of binary star: a polar with an extremely strong magnetic field. The magnetic field around the white dwarf star is so powerful that it distorts the main sequence star that is its binary partner into an egg-shaped configuration and synchronizes the orbit of the system so that the same side of the star always faces the white dwarf. AM Herculis is a highly energetic and cataclysmic variable.

### What is a **cataclysmic variable**?

A cataclysmic variable is a binary star system that periodically has a huge explosion at the surface of one of the stars. Most often, the cataclysmic variable consists of a white dwarf and a main sequence star. Matter from the larger, more distended main sequence star flows down toward the surface of the white dwarf. When the accreted material reaches a certain critical mass, it detonates in a powerful thermonuclear explosion. The star is not destroyed, though. After this big flare-up, the cycle of accretion and explosion occurs again, sometimes after a few hours, and sometimes after a few centuries.

## Does the Sun have a binary companion?

**A**lthough a binary companion to the Sun has never been detected, it is remotely possible that a very faint, very distant star could be orbiting the solar system at a great distance, similar to the way that Proxima Centauri (Alpha Centauri C) may be orbiting Alpha Centauri A and B. This idea has been explored in popular science fiction, and this tiny companion has been nicknamed Nemesis, the avenging goddess of justice in ancient Greek mythology, who was also called the Daughter of the Night. Some people hypothesize that such a companion might occasionally change the orbit of distant comets just enough that they might plunge in toward the center of our solar system and strike Earth. There is no scientific evidence to support these ideas, however.

One particular kind of cataclysmic variable is called a classical nova. This is not to be confused with a supernova, which is an explosion that obliterates a star. Still, even though classical novae are not quite as titanic, they are very powerful and impressive.

## What is a **Cepheid variable**?

Cepheid variable stars are not binary stars as cataclysmic variables are. Rather, they are single stars that pulsate—grow and shrink in size, with a corresponding change in their luminosity—because of internal processes. Cepheid variables have played a key role in the study of the universe, because their pulsations create a period-luminosity relation that allows them to be used as standard candles for distance determinations.

## What is an **RR Lyrae star**?

An RR Lyrae star, like a Cepheid variable, also pulsates because of internal processes. It too follows a period-luminosity relation, and can be used as a standard candle. In fact, RR Lyrae stars were used as standard candles before Cepheids; they helped astronomers determine the size of the Milky Way galaxy by measuring the distances to star clusters that orbit the center of our galaxy. RR Lyraes are not as famous as Cepheids as standard candles, mostly because they are somewhat fainter than Cepheids and are not easily usable at very large (intergalactic) distances. Still, they are uniquely valuable because they are much older than Cepheids, so they can be used as distance indicators to objects with older stellar populations.

# STAR CLUSTERS

## What is a **star cluster**?

Stars are often grouped together in space. These groupings are called star clusters, and they are different from constellations in that they are actually physically associated with one another, rather than just appearing that way. The best-known kinds of star clusters are globular clusters and open clusters.

## How are star clusters formed?

Current theory and observations suggest that clusters almost always form from a single, very large cloud of gas. All of the stars in the cluster form over the same short period of time (anywhere from a few thousand to a few million years). Open clusters are fairly young structures and usually dissipate from the random motions of the stars after a few hundred million years or a few billion years at most. Globular clusters stick together tightly, by contrast, and can last for many billions of years.

## What is an open cluster?

Open clusters form quickly and often; they are much smaller than globular clusters. They usually contain a few dozen to a few hundred stars; and they do not form into any particular shape. As the name implies, they look more irregular and open.

## How many open clusters are there?

In our Milky Way galaxy, more than 1,000 open clusters have been found. There may also be many more open clusters hidden from view by clouds of dusty gas within the galaxy.

## What are some examples of well-known open clusters?

In the southern hemisphere, the Jewel Box is a particularly beautiful open cluster that looks as if it contains sparkling stars with several different colors. In the northern hemisphere, the Hyades (also known as the Beehive) is a well-known open cluster; slightly to the east of the Hyades, in the direction of the constellation Taurus the Bull, is probably the best-known open cluster in the night sky: the Pleiades, or Seven Sisters.

## What are some of the best-known star clusters?

The following table lists some of the better known clusters.

**Some Well-Known Star Clusters**

| Name | Catalog Name | Cluster Type |
| --- | --- | --- |
| 47 Tucanae | NGC 104 | globular cluster |
| Beehive Cluster | Messier 44 | open cluster |
| Christmas Tree Cluster | OC NGC 2264 | open cluster |
| Hercules Cluster | Messier 13 | globular cluster |
| Hyades | Melotte 25 | open cluster |
| Jewel Box | NGC 4755 | open cluster |
| Messier 3 | NGC 5272 | globular cluster |
| Omega Centauri | NGC 5139 | globular cluster |
| Pleiades | Messier 45 | open cluster |
| Trapezium Cluster | Orion Trapezium | embedded cluster |

## What is—or what are—**the Pleiades**?

The Pleiades is an open cluster about four hundred light-years from Earth. It contains dozens of stars, of which the six or seven brightest (named Alcyone, Atlas, Electra, Maia, Merope, Taygeta, and Pleione) are readily visible to the naked eye. The stars are embedded in a small but bright reflection nebula, which makes this open cluster particularly easy to see. The Pleiades is also known as the Seven Sisters in Europe and America, and has inspired the sky-lore and legend of many ancient cultures.

## How did **ancient people** use the **Pleiades** to mark **seasons** and **calendar** cycles?

In many ancient cultures, the Pleiades was associated with the changing of the seasons. That is because in Earth's northern hemisphere the Pleiades becomes visible in the sky at dawn in the spring and at sunset in the fall. This led it to be a symbol of the times of sowing and harvest. The ancient Aztecs of Mexico based their 52-year calendar cycle on the position of the Pleiades. They began each new cycle when the Pleiades ascended to a position directly overhead at the sky's zenith. At midnight on that day, the Aztecs performed an elaborate ritual celebrating the heavens and the earth.

## What is a **globular cluster**?

Globular clusters are nearly spherical distributions of stars, usually a few dozen to a few hundred light-years across. They contain anywhere from several thousand to several million stars, and they are packed relatively close together. The stars

The Pleiades. (*NASA/JPL-Caltech/J. Stauffer*)

## What are some interesting myths behind the Pleiades?

According to one story in Greek mythology, the Pleiades were Pleione, the wife of Atlas the Titan (who supported Earth on his shoulders as punishment for turning against the gods), and their daughters. The Pleiades were being pursued by the hunter Orion, and Zeus helped them escape. He first turned them into doves, and they flew away from Orion; then Zeus lifted them into the sky as stars.

On the other side of the globe, there is an Australian aboriginal folktale about the Pleiades that portrays the stars as a group of women being chased by a man named Kulu. Two lizard men, together known as Wati-kutjara, came to the rescue of the women. They threw their boomerangs at Kulu and killed him. The blood drained from Kulu's face, and he turned white and rose up into the sky to become the Moon. The lizard men became the constellation Gemini, and the women turned into the Pleiades.

are held together by their mutual gravity and are most heavily concentrated at the center of the cluster. In at least one instance—the cluster G1 that orbits the Andromeda galaxy—there appears to be a black hole at the center of a globular cluster.

## How **many globular clusters** are there?

Every large galaxy has its own system of globular clusters. Around the Milky Way, there are between 150 and 200 globular clusters. There are about twice that number orbiting the Andromeda galaxy, our nearest large galaxy neighbor. Around some large elliptical galaxies, thousands of globular clusters have been detected.

M80 is a globular star cluster located about 28,000 light-years from Earth and containing hundreds of thousands of stars. (*NASA, The Hubble Heritage Team, STScI, AURA*)

### How **old** can **globular clusters** get?

Current astronomical evidence suggests that some globular clusters may have been the oldest stellar collections to form early in the history of the universe. By studying the color-magnitude diagrams of globular clusters, astronomers have concluded that some of them are at least 12 billion years old, which is as old as the most distant galaxies yet observed.

### What are some **examples** of **well-known globular clusters**?

In the northern hemisphere, the Hercules Cluster is easily visible with binoculars or small telescopes. In the southern hemisphere, two prominent globular clusters are easily visible on a dark night with the unaided eye: 47 Tucanae and Omega Centauri.

## What is the **difference** between a **large star cluster** and a **small galaxy**?

Astronomers have been trying to answer this question for many years. Omega Centauri, for example, contains several million stars, as does 47 Tucanae. Many dwarf galaxies have comparable numbers of stars, so it is not perfectly clear where a "star cluster" ends and a "galaxy" begins, when it comes to classifying stellar collections of this size. There may be differences with regard to diameters, or perhaps dark matter content, that will eventually help astronomers find a definitive distinction between these two kinds of of objects.

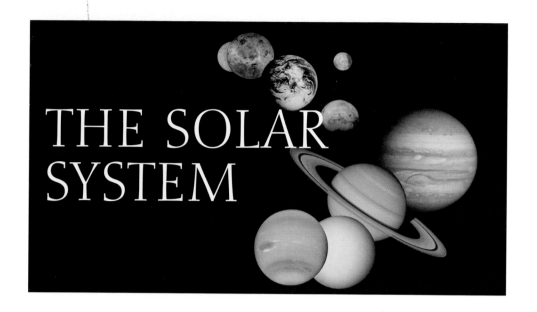

# THE SOLAR SYSTEM

## PLANETARY SYSTEMS

### What is a **planetary system**?

A planetary system is a system of astronomical objects that populate the vicinity of a star. This includes objects like planets, asteroids, comets, and interplanetary dust. In a more general sense, this also includes the star itself, its magnetic field, its stellar wind, and the physical effects of those things, including ionization boundaries, and shock fronts.

### What is **our own planetary system** called?

The Sun is the gravitational anchor of the planetary system where we live. The term "solar" refers to anything having to do with the Sun; so we call our own planetary system the solar system. Often, astronomers will refer to other planetary systems as "solar systems" too, though that is not technically correct.

### How did the **solar system form**?

The solar system probably formed in a way that follows the basic ideas of the so-called nebular hypothesis, which was advanced in the eighteenth century by Pierre-Simon de Laplace (1749–1827) and significantly updated since that time. About 4.6 billion years ago, the Sun formed from a large cloud of gas and dust that collapsed upon itself because of gravitational instability. When the Sun was born, not all of the nebula of gas and dust that had been gravitationally gathered was incorporated into the Sun itself. Some of it settled into a disk of orbiting material. As this material orbited in a protoplanetary disk, numerous collisions between the tiny grains led to some of the grains sticking together, making larger bodies. After millions of years, the largest bodies—planetesimals—had sufficient mass (and hence gravity)

In this artist's depiction, a star is surrounded by matter that will eventually combine to form orbiting planets. (*NASA/JPL-Caltech/T. Pyle*)

to start attracting other objects in the disk to them. Growing larger and larger, these planetesimals became protoplanets; the largest protoplanets grew larger still, until at last the planets were formed. Although the solar wind has removed much of the remaining, unprocessed gas and dust, numerous smaller objects (and some of the gas and dust, as well) still remain today, providing the rich variety of objects and phenomena in a solar system more than four and a half billion years later.

### How **large** is our **solar system**?

Our solar system reaches out to the orbit of the most distant planet, Neptune, or about three billion miles (five billion kilometers) away from the Sun. Beyond Neptune is the Kuiper Belt, a thick, doughnut-shaped cloud of small icy bodies that extends to about eight billion miles (12 billion kilometers). Beyond that still is the Oort Cloud, which is a huge, thick, spherical shell thought to contain trillions of comets and comet-like bodies. The Oort Cloud may extend as far as a light-year, nearly six trillion miles, out from the Sun.

### What is the scientific **origin** of the **nebular hypothesis** of solar system formation?

The original nebular hypothesis was first suggested around 1755 by the German philosopher Immanuel Kant (1724–1804), and later advanced by the French mathe-

## Are there planetesimals and protoplanets outside our solar system?

Since there are more than 200 confirmed planets that exist beyond our solar system and orbit other stars, it is likely that planetesimals and protoplanets exist beyond our solar system as well. Such objects no doubt populate protoplanetary nebulae around other stars. One well-known example is the disk of gas and dust around the star Beta Pictoris. Observations with infrared telescopes, such as the Infrared Astronomical Satellite (IRAS) and the Spitzer Space Telescope, have further detected dozens of stars surrounded by cocoons of dense dusty gas, where protoplanetary accretions are most likely taking place.

matician and scientist Pierre-Simon de Laplace. The idea was similar to the current theory of the formation of the Sun, but differed in the way that planets supposedly formed. Laplace suggested that the Sun formed a spinning nebula, and that as the nebula contracted toward the Sun it gave off rings of gas. Material in these orbiting rings then condensed into the planets through collisions and gravitational attraction. This version of the nebular hypothesis was published in Laplace's 1796 book, *Exposition du Systeme du Monde* (*The System of the World*). Although it was not correct in its details, it was a strong pioneering effort in pursuing our astrophysical origins.

## What are **planetesimals**?

Planetesimals are early solar system objects that range in size from about 0.6–60 miles (1–100 kilometers) across. Like so many terms in science, this is not an exact definition. More generally, it refers to objects in the protoplanetary nebula that have formed by collisions and may be starting to accrete more material via their gravitational influence.

## What are **protoplanets**?

Protoplanets are early solar system objects that range in size from about 60–6,000 miles (100–10,000 kilometers) across. Again, like the term "planetesimal" and many other terms in science, this is not an exact definition. More generally, protoplanets are objects in the protoplanetary nebula that are large enough that they are growing in size and mass by attracting other, smaller objects with their gravitational pull.

## What are the **major zones** of the **solar system**?

Scientists generally divide the solar system into five major zones: the inner (or terrestrial) planet zone, the asteroid belt, the outer (or gas giant) planet zone, the Kuiper Belt, and the Oort Cloud. There is no exact boundary for these zones, however, and their sizes are not well determined; there is also overlap, in the sense that objects from one zone often appear in another zone.

# PLANET BASICS

### What is a **planet**?

There have been many attempts to define the term "planet" over the centuries, but to date there is still no universally agreed-upon scientific definition of the term. Generally speaking, however, a planet usually refers to an object that is not a star (that is, has no nuclear fusion going on in its core); that moves in orbit around a star; and is mostly round because its own gravitational pull has shaped it into, more or less, a sphere.

Our solar system officially contains eight planets, including (clockwise from top left), Mercury, Venus, Earth (shown with Moon), Mars, Jupiter, Saturn, Uranus, and Neptune. (*NASA*)

### What are the **general characteristics** of the **planets** in our **solar system**?

All the planets in our solar system, by the current scientific classification system, must satisfy three basic criteria:

1. A planet must be in hydrostatic equilibrium—a balance between the inward pull of gravity and the outward push of the supporting structure. Objects in this kind of equilibrium are almost always spherical or very close to it.

2. A planet's primary orbit must be around the Sun. That means objects like the Moon, Titan, or Ganymede, are not planets, even though they are round due to hydrostatic equilibrium, because their primary orbit is around a planet.

3. A planet must have cleared out other, smaller objects in its orbital path, and thus must be by far the largest object in its orbital neighborhood. This means that Pluto is not a planet, even though it meets the other two criteria; there are thousands of Plutinos in the orbital path of Pluto, and it crosses the orbit of Neptune, which is a much larger and more massive object.

The eight objects in our solar system that meet all three criteria are Neptune, Uranus, Saturn, Jupiter, Mars, Earth, Venus, and Mercury.

### What are the **masses, orbital periods,** and **positions** of the **planets** in our solar system?

The table below lists the basic information about the planets in our solar system.

## Who decides what is a planet and what is not?

For about two centuries, the International Astronomical Union has been the official standards-governing body of professional astronomers worldwide. Official names of objects in the universe—for example, asteroids or comets or planets—are suggested to, then approved or rejected by the IAU committee on names and nomenclatures. The IAU formed a special committee to decide how to classify planets in our solar system because it was becoming scientifically clear that Pluto and other Kuiper Belt Objects would have to be designated in a scientifically valid, practically sensible way.

### The Planets of Our Solar System

| Name | Mass (in Earth masses*) | Diameter (in Earth diameters**) | Distance to Sun (in AU***) | Orbital Period (in Earth years) |
|---|---|---|---|---|
| Mercury | 0.0553 | 0.383 | 0.387 | 0.241 |
| Venus | 0.815 | 0.949 | 0.723 | 0.615 |
| Earth | 1 | 1 | 1 | 1 |
| Mars | 0.107 | 0.532 | 1.52 | 1.88 |
| Jupiter | 317.8 | 11.21 | 5.20 | 11.9 |
| Saturn | 95.2 | 9.45 | 9.58 | 29.4 |
| Uranus | 14.5 | 4.01 | 19.20 | 83.7 |
| Neptune | 17.1 | 3.88 | 30.05 | 163.7 |

*One Earth Mass equals $5.98 \times 10^{24}$ kilograms.

**One Earth Diameter equals 12,756 kilometers.

***An astronomical unit (AU) is the distance from Earth to the Sun and is roughly $1.5 \times 10^8$ kilometers.

## What is the current, **official planetary classification system**?

On August 24, 2006, the general assembly of the International Astronomical Union approved the current system of classifying planets in our solar system. This system added a specific scientific requirement for planethood: it must have cleared all other significantly sized bodies out of its orbital path or neighborhood, probably through collisions or gravitational interactions. This system also creates a new designation called a "dwarf planet," which describes an object that fulfills all the criteria of a planet except this one. This system, like every other classification that has come before it, has strengths and weaknesses; no matter what, though, it gives all people a starting point to learn about—and hopefully understand—what planets are all about.

This current classification system means that, officially, there are eight planets in the solar system—Mercury, Venus, Earth, Mars, Jupiter, Saturn, Uranus, and Neptune—and a number of dwarf planets, including Pluto, Charon, Ceres, Eris, and Quaoar.

## What was the **previous planetary classification system**?

The previous classification system was based on historical knowledge and size. The eight planets in our solar system today, plus Pluto, were known to scientists and were believed to be large—at least, all larger than Earth's moon. Other objects that were known to orbit the Sun as its primary, but were smaller than about 2,000 miles across, were called asteroids (or, more generally, minor planets). So until August 24, 2006, the International Astronomical Union officially recognized Pluto as the ninth planet. It does not anymore.

## Have **planetary reclassifications** occurred **before Pluto's**?

Yes, and one is bound to occur again someday. In ancient times, the term planet meant any object that moved naturally across the sky compared with the background view of the stars; that meant the planets included the Sun, Moon, Mercury, Venus, Mars, Jupiter, and Saturn. Over time, the planet Uranus was discovered in the late 1700s, and the Sun and Moon were dropped from the category. In the 1800s, nearly a dozen small objects that orbited the Sun were declared to be planets; they were then reclassified as asteroids, but Neptune was not and remained a planet. Pluto is just the latest recategorization in a long history.

## What are some kinds of **unofficial classifications** of **planets**?

Unofficial classifications of planets in our own solar system include terrestrial planets, gas giant planets, major planets, minor planets, inner planets, outer planets, and possibly icy planets. Remember, though, that more than 200 planets outside our solar system are now known. So other unofficial categories like exoplanets, hot Jupiters, and rogue planets are now used as well.

## What is a **planetary ring**?

A planetary ring is a system of huge numbers of small bodies—ranging in size from grains of sand to house-sized boulders—that orbit in a coherent ring-shaped pattern around a planet. The most spectacular planetary rings in the solar system orbit around Saturn; they are more than 170,000 miles across, and are less than one mile thick.

# THE INNER SOLAR SYSTEM

## What **planets** are included in the **inner solar system**?

The planets that are collectively thought of as belonging to the inner solar system are Mercury, Venus, Earth, and Mars.

## What is the **terrestrial planet zone** and what does it contain?

The terrestrial planet zone is generally considered to be the part of the solar system containing the planets Mercury, Venus, Earth and Mars. These four objects are called the terrestrial planets because they resemble one another (specifically, Earth) in their structure: a metallic core, surrounded by a rocky mantle and thin crust. There are three moons in the terrestrial zone as well: Earth's moon, and the two moons of Mars: Phobos and Deimos.

## What are the **physical properties** of **Mercury**?

Mercury's diameter is a little more than one-third that of Earth's, and it has just 5.5 percent of Earth's mass. On average, Mercury is 58 million kilometers (36 million miles) away from the Sun. That is so close to the Sun that Mercury's orbit is rather tilted and stretched into a long elliptical shape. Mercury orbits the Sun in just 88 Earth days, but Mercury's day—the time it takes to rotate once around its polar axis—is about 59 Earth days.

A 1974 image of Mercury taken by a camera on *Mariner 10*. (*NASA*)

Mercury's surface is covered with deep craters, separated by plains and huge banks of cliffs. There is absolutely no water on the planet. Mercury's most notable surface feature is an ancient crater called the Caloris Basin, which is about five times the size of New England—a huge pit for such a small planet! Mercury's very thin atmosphere is made primarily of sodium, potassium, helium, and hydrogen. On its day side (the side facing the Sun), temperatures reach 800 degrees Fahrenheit (430 degrees Celsius); on its night side, the heat escapes through the negligible atmosphere, and temperatures plunge to 280 degrees below zero Fahrenheit (–170 degrees Celsius).

## Is it easy to **see Mercury** from Earth?

Since Mercury is so close to the Sun, the glare of the Sun makes it difficult to observe Mercury from Earth. Mercury is therefore visible only periodically, when it is just above the horizon, for at most an hour or so before sunrise and after sunset. It also moves more quickly across the sky than the other planets. Even when Mercury is visible, the sky is often so bright that it is hard to distinguish it from the background sky.

## What is **Mercury's history**?

Astronomers think that Mercury, like the moon, was originally made of liquid rock and that the rock solidified as the planet cooled. Some meteorites hit the planet during the cooling stage and formed craters. Other meteorites, however, were able to break through the cooling crust. The impact caused lava to flow up to the surface and cover over older craters, forming the plains.

## What are the **physical properties** of **Venus**?

Venus is similar to Earth in many ways. Venus is closer in distance to Earth than any other planet, and it has a similar size and composition. However, the surface characteristics differ greatly.

A year on Venus is equal to 225 Earth days, compared to Earth's 365-day year. Venus, however, rotates on its polar axis backwards compared to Earth, so a Venus sunrise occurs in the west and sunset is in the east. Furthermore, a Venusian day is 243 Earth days long, which makes it even longer than a Venusian year.

The surface conditions of Venus are far different from that of our own planet. It is blanketed by a thick atmosphere nearly 100 times denser than ours, and it is made mostly of carbon dioxide, along with some nitrogen and trace amounts of water vapor, acids, and heavy metals. Venus's clouds are laced with poisonous sulfur dioxide, and its surface temperature is a brutal 900 degrees Fahrenheit (500 degrees Celsius). Interestingly, this is even hotter than Mercury, which is much closer to the Sun. These hostile conditions came about because of a runaway greenhouse effect on Venus that persists to this day.

## What is the **runaway greenhouse effect**?

On Venus, the greenhouse effect "ran away." The heat trapped by the Venusian atmosphere caused the surface temperature to get so high that the rocky crust began to release greenhouse gases like carbon dioxide. The atmospheric insulation consequently became even thicker, which caused more heat to be trapped, which caused the temperature to rise higher still, which caused even more greenhouse gas to be released. After finally reaching thermal equilibrium, Venus is now the inferno we see today.

## What is the **surface of Venus** like?

Venus appears to have a rocky surface covered with volcanoes, some of which may still be active. Volcanic features such as lava plains, channels that look like dry riverbeds, mountains, and medium and large craters can also be found. No small craters exist; this may be because small meteorites cannot penetrate the planet's thick atmosphere to strike the surface and make a crater. One particularly interesting set of features found on the surface are called arachnoids. These are circular geological formations ranging from 30 to 140 miles (50 to 220 kilometers) across that are filled with concentric circles and "spokes" extending outward.

## What is the greenhouse effect?

**A**s its name implies, the greenhouse effect occurs on planets with atmospheres, and causes a planet's surface to be warmer than it would be without that atmosphere. In a greenhouse on Earth, transparent glass walls, doors, and roofs let visible sunlight in, which then strikes the objects inside the greenhouse and converts into heat. The heat tries to escape as invisible infrared radiation, but the glass blocks the infrared light. Heat builds up inside the greenhouse, and its temperature is much warmer than the air outside. When the greenhouse effect happens on a planet, gases in that planet's atmosphere block infrared radiation from leaving the planet's surface, much like the glass on a greenhouse. Carbon dioxide and water vapor are common gases that trap heat very effectively; so planets with thick atmospheres containing large amounts of these gases can get much warmer than they otherwise would be.

Maps of Venus made by the *Magellan* orbiter showed that, in a geologic sense, the Venusian surface is relatively young. Not long ago, lava appears to have erupted from some source and covered the entire planet, giving it a fresh, new face. One piece of evidence that supports this hypothesis is that there are craters and other geologic formations that lack the weathered, worn appearance expected of older features. Also, there are surprisingly few craters on Venus for a planet of its size. In fact, more craters can be counted when viewing a section of Earth's moon through a small telescope than occur on the entire surface of Venus.

## What does **Venus look like** from **Earth**?

Since Venus is closer to the Sun than Earth, it is never up in the sky at midnight. Rather, Venus is visible in the sky either just after dark or just before sunrise, depending on the season. (This pattern of appearance prompted ancient astronomers to refer to Venus as the "evening star" or the "morning star.")

Due to its proximity to Earth, and to the highly reflective cloud layers in its atmosphere, Venus can look incredibly bright and beautiful in the sky. At its brightest, it is the third brightest object in the sky, after the Sun and the Moon. Like the Moon, Venus is often visible in the daytime, as long as one knows where to look. It is no wonder that Venus is named for the Roman goddess of love and beauty.

Through a small telescope, it is possible to see Venus undergo phases, just like the Moon. This occurs because, from our point of view on Earth, we see only the parts of Venus that are illuminated by sunlight at any given time. Unlike the Moon, though, Venus is usually brighter to our view in its crescent phase than in its full phase. At its brightest, Venus is the object in the night sky most likely to be mistaken for an aircraft or a UFO.

### What are the **physical properties** of **Mars**?

Mars is the fourth planet from the Sun in our solar system. Its diameter is about half that of Earth, and its year is about 687 Earth days. That means that its seasons are about twice as long as ours here on Earth. However, a Martian day is very close in length to an Earth day—only about 20 minutes longer, in fact.

The Martian atmosphere is very thin—only about seven-thousandths the density of Earth's atmosphere. The atmosphere is mostly carbon dioxide, with tiny fractions of oxygen, nitrogen, and other gases. At the equator, during the warmest times of the Martian summer, the temperature can reach nearly zero degrees Fahrenheit (–18 degrees Celsius); at the poles, during the coldest times of the Martian winter, temperatures drop to 120 degrees below zero Fahrenheit (–85 degrees Celsius) and beyond.

Mars has fascinating geologic features on its surface; it is covered with all sorts of mountains, craters, channels, canyons, highlands, lowlands, and even polar ice caps. Scientific evidence strongly suggests that once, billions of years ago, Mars was much warmer than it is now, and was an active, dynamic planet.

### Who **discovered** the **polar ice caps** on **Mars**?

The Italian astronomer Gian Domenico Cassini (1625–1712) made a number of important discoveries, including a gap in Saturn's rings (known today as the Cassini division). He made detailed observations of Mars, and discovered light-colored patches at the Martian north and south poles. These polar caps showed seasonal variations, spreading during the Martian winter and shrinking during the summer.

### What are the **Martian polar ice** caps **made of**?

Current studies suggest that the Martian polar ice caps are made up mostly of frozen carbon dioxide, also known as "dry ice." Some frozen water, or just plain ice, may also be imbedded within the polar caps. Due to the atmospheric conditions on the surface of Mars, however, neither the ice nor the dry ice would melt to make water or liquid carbon dioxide when the temperatures go up; rather, they would sublimate, or turn directly into gas. So, unlike here on Earth, the polar ice caps on Mars are not a source of liquid water.

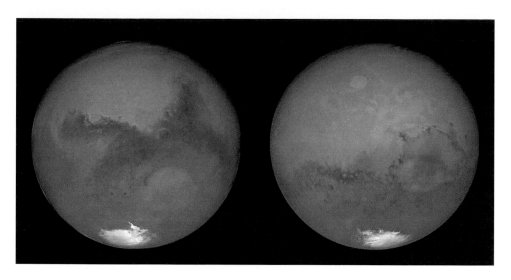

In this image from 2003, the two sides of Mars are shown, with Olympus Mons clearly visible in the right image on the northern portion of the planet. The southern ice cap is also visible. (*NASA and Space Telescope Science Institute*)

## What are some of the most **interesting geological features** of **Mars**?

Mars has a rich variety of geological features: huge craters, broad plains, tall mountains, deep canyons, and much more, all with colorful names. The tallest mountain in the solar system, the extinct volcano Olympus Mons, rises 15 miles (24 kilometers) above the Martian surface. A massive canyon called the Vallis Marineris (Mariner Valley) cuts across the northern hemisphere of Mars for more than 2,000 miles (3,200 kilometers); it is three times deeper than the Grand Canyon. On Earth, the Vallis Marineris would stretch from Arizona to New York. A noteworthy feature on the southern hemisphere of Mars is Hellas, an ancient canyon that was probably filled with lava long ago and is now a large, light area covered with dust.

## What is some of the **geological history** of **Mars**?

Mars was almost certainly much warmer billions of years ago than it is today. Water may have once flowed across the Martian surface the way rivers and streams flow across Earth's surface today. There were probably alluvial plains, deltas, lakes, and perhaps even seas and oceans, too. The internal heat under the Martian crust probably powered volcanism and massive magma and lava flows. Furthermore, since the gravitational pull at the Martian surface is about one-third that of Earth's, volcanic cones and other mountains could be built higher than on Earth, and canyons cut deeper because landslides and erosion would not be as strong an influence.

## How do we know there was **once liquid water** on **Mars**?

Orbital data shows features clearly attributable to flowing liquids: riverbeds, tributary structures, and deltas leading to low-altitude areas, for example. From the sides of some steep craters, images show tracks as if water had burst through the crust, then flowed out, and then froze or evaporated.

135

## What is the story behind the Martian meteorite ALH84001?

**A**LH84001 was so named because it was found in the Allan Hills region of Antarctica in 1984 by Roberta Score, a member of the Antarctic Search for Meteorites (ANSMET) team. It is the most famous of a number of meteorites that are thought to have been pieces of the Martian surface millions of years ago. They were probably knocked loose by a powerful collision from a comet or asteroid, which sent pieces of rock into orbit around the Sun that later landed on Earth.

Several kinds of scientific evidence are used to determine where meteorites come from. These include the crystallization age of the meteorite, its chemical and physical composition, the effects that cosmic rays have had on it, and the composition and concentrations of gases trapped long ago in tiny fissures and bubbles in the meteorite. From this evidence it was determined that ALH84001 originated on Mars.

In 2005 additional evidence was shown that suggests the existence of a vast frozen sea of water ice below the surface. Doppler mapping technology on Mars orbiters—similar to those used by weather satellites orbiting Earth, but adapted for underground investigation—was used to find a body of ice, ranging in depth of a few feet to several hundred feet, that covered an area larger than the states of New York, New Jersey, Pennsylvania, Ohio, and Indiana combined.

## What **evidence** have we gathered from the **Martian surface** to show there was once **liquid water** on Mars?

The Mars Exploration Rovers, *Spirit* and *Opportunity,* are geological robots that have explored several areas of Mars. Among the many discoveries made with them are minerals that form only in the long term presence of water; microscopic mineral structures nicknamed "blueberries" that only form when moisture is present, along with chemical and isotopic ratios in Martian rocks that would have formed only if liquid water were in the environment. The strong scientific conclusion is that Mars is currently dry on it surface, but that this was not always the case. It may even have been awash with liquid water billions of years ago.

# GAS GIANTS

## What is a **gas giant planet**?

Gas giant planets are so named because they are much larger than the terrestrial planets, and they have atmospheres so thick that the gas is a dominant part of the planets' structure.

*Voyager 2* took these images of the gas giant planets (from left to right) Neptune, Uranus, Saturn, and Jupiter. *(NASA)*

### Which planets in our solar system are considered gas giants?

Jupiter, Saturn, Uranus, and Neptune are all categorized as gas giants.

### What is the gas giant zone?

The gas giant zone is the part of the solar system roughly between the orbit of Jupiter and the orbit of Pluto. It contains the outer (gas giant) planets Jupiter, Saturn, Uranus, and Neptune. Each of the gas giant planets has a host of moons and rings or ringlets.

### What are the physical properties of Jupiter?

Jupiter is by far the largest planet in our solar system. It is about twice as massive as all the other planets, moons, and asteroids in our solar system put together. However, its day is only 10 hours long, less than half an Earth day. The fifth planet out from the Sun, Jupiter is 1,300 times Earth's volume and 320 times Earth's mass.

More than 90 percent of Jupiter's mass consists of swirling gases, mostly hydrogen and helium; in this incredibly thick, dense atmosphere, storms of incredible magnitude rage and swirl. The largest of these storms is the Great Red Spot, which is often visible from Earth through even a small telescope.

Jupiter has a rocky core made of material thought to be similar to Earth's crust and mantle. However, this core may be the size of our entire planet, and its temperature may be as high as 18,000 degrees Fahrenheit (10,000 degrees Celsius), with pressures equal to two million Earth atmospheres. Around this core, in these

137

In an image taken from *Voyager 1* Jupiter's Great Red Spot can be seen with the various cloud colors. (*NASA*)

extreme conditions, it is likely that a thick layer of compressed hydrogen is present; the hydrogen in this layer probably acts like metal, and may be the cause of Jupiter's intense magnetic field, which is five times greater than even that of the Sun.

At least 30 moons orbit Jupiter. Many of them are only a few miles across and are probably captured asteroids. However, four of them—Io, Europa, Ganymede, and Callisto—are about the size of Earth's Moon or larger.

## What are some other **characteristics** of **Jupiter's atmosphere**?

The mini-probe launched in 1995 from the *Galileo* spacecraft made detailed measurements of Jupiter's atmosphere down to about 90 miles (150 kilometers) below the cloud-tops. It found that these upper layers of Jupiter's deep, dense atmosphere contain water vapor, helium, hydrogen, carbon, sulfur, and neon, all in lower concentrations than were previously predicted. On the other hand, it had higher concentrations of other gases, such as krypton and xenon.

Scientists were also surprised by what the probe did not find. Rather than several dense cloud layers of ammonia, hydrogen sulfide, and water vapor, as was predicted, the probe only detected thin, hazy clouds. Also, scientists had predicted tremendous amounts of lightning discharges; but only faint signs of lightning at least 600 miles (1,000 kilometers) away were detected. This suggested that, at these atmospher-

## What do we know about Jupiter's Great Red Spot?

The Great Red Spot is a huge windstorm more than 8,500 miles (14,000 kilometers) wide and 16,000 miles (26,000 kilometers) long. You could easily place the planets Earth and Venus side-by-side inside the Great Red Spot! The storm that perpetuates the Spot is apparently powered by the upswell of hot, energetic gases from deep inside Jupiter's atmosphere, which produce winds that blow counterclockwise around the Spot at 250 miles (400 kilometers) per hour.

The Great Red Spot may derive its red color from sulfur or phosphorus, but this has not been conclusively shown. Beneath it are three white, oval areas; each is a storm about the size of the planet Mars. There are thousands of huge and powerful storms on Jupiter, and many of them can last for a very long time. However, the Great Red Spot, which has been going on for at least 400 years, and which was first studied by Galileo Galilei, remains the biggest and most visible Jovian storm yet recorded.

ic depths, lightning occurs on Jupiter only about one-tenth as often as it does on Earth. It is important to note, though, that the surprising results from *Galileo*'s miniprobe were only obtained from one area in Jupiter's atmosphere. It is possible that the atmospheric conditions there were not representative of the entire atmosphere.

## How was **Jupiter formed**?

Jupiter is the archetypal gas giant planet—so much so that gas giants are often called Jovian planets. Thus, Jupiter is thought to have formed pretty much the same way that all other gas giants form. Although many details remain uncertain, scientists think that Jupiter formed soon after the Sun itself. As the solar nebula settled into a swirling disk of dust and gas, small particles came together over millions of years' time and eventually formed planetesimals, which in turn came together to form the core of Jupiter. The planet's core then attracted the gas in and around its orbital path, which gathered and coalesced into Jupiter's massive atmosphere.

## Who first **measured Jupiter's size**?

In 1733 English astronomer James Bradley (1693–1762) succeeded in measuring Jupiter's diameter, shocking the scientific community at the time with the news of the planet's immense size.

## Does **Jupiter** have a **magnetic field**?

Yes, and it is about five times the intensity of the Sun's. Jupiter's magnetosphere is so big that it would take up a good part of our night sky—much larger than the full

Moon—if we could see it with our eyes. Also, like Earth, there are large belts of trapped, highly energized charged particles around Jupiter; these "van Allen" belts are confined by lines of magnetic force that have naturally developed in Jupiter's magnetic field.

## Does **Jupiter** have **rings**?

Yes, Jupiter has several very faint rings. They are nothing like Saturn's enormously developed and beautiful rings, but they can be detected through careful observations with instruments like the Hubble Space Telescope.

## What are the **physical properties** of **Saturn**?

Saturn is similar to Jupiter, though about one-third the mass. Still, it is about 95 times more massive than Earth. Saturn's average density is actually lower than that of water. A day on Saturn is only 10 hours and 39 minutes long; it spins so fast that its diameter at the equator is 10 percent larger than its diameter from pole to pole.

Saturn has a solid core likely made of rock and ice, which is thought to be many times the mass of Earth. Covering this core is a layer of liquid metallic hydrogen, and on top of that are layers of liquid hydrogen and helium. These layers conduct strong electric currents that, in turn, generate Saturn's powerful magnetic field.

Saturn has dozens of moons, and its largest moon is Titan, which is larger than Earth's own moon and has a thick, opaque atmosphere. The most spectacular part of Saturn is its magnificent system of planetary rings, which stretch some 170,000 miles (300,000 kilometers) across.

## What is **Saturn's atmosphere** like?

Saturn has hazy, yellow cloud-tops made primarily of crystallized ammonia. The clouds are swept into bands by fierce easterly winds that have been clocked at more than 1,100 miles (1,800 kilometers) per hour at the equator. Saturn's winds near its poles are much tamer. Also like Jupiter, powerful cyclonic storms appear on Saturn often. About every 30 years, for example, a massive storm forms that appears white. Known as the "Great White Spot"—even though it is not the same storm every time—it can be visible for up to a month, shining like a spotlight on the planet's face, before it dissipates and stretches around the planet as a thick white stripe. This recurring storm is thought to be a result of the warming of Saturn's atmosphere toward the end of the Saturnian summer, which causes ammonia deep inside the atmosphere to bubble up to the cloud-tops, only to be whipped around by the planet's powerful winds.

While other planets in the solar system have rings, Saturn's are easily the most stunning, as seen in this image from *Voyager 2*. *(NASA)*

## What are **Saturn's rings** like?

Saturn's ring system is divided into three main parts: the bright A and B rings and the dimmer C ring. (There are many other fainter rings as well.) The A and B rings are divided by a large gap called the Cassini Division, named after Gian Domenico Cassini (1625–1712). Within the A ring itself is another division, called the Encke Gap after Johann Encke (1791–1865), who first found it in 1837. Although these gaps appear to be completely empty, they are nonetheless filled with tiny particles, and, in the case of the Cassini Division, dozens of tiny ringlets.

Although Saturn's rings measure more than 100,000 miles across, they are only about a mile or so (one or two kilometers) thick. That is why they sometimes seem to disappear from view here on Earth. When the orbit of Saturn is such that we see the rings edge-on, the rings look like a thin line and can be nearly invisible.

## Who **discovered Saturn's rings**?

Galileo Galilei (1564–1642) first observed Saturn's rings, but he could not figure out what they were. To him the rings looked like "handles." He communicated his discovery to other scientists in Europe, one of whom was the Dutch scientist Christian Huygens (1629–1695). Using his own telescopes, Huygens found that these handles, which looked like moons on either side of Saturn, were actually parts of a large disklike ring. Huygens continued to study Saturn over a long period of time, showing how the changing angle of the planet's tilt caused the ring's changing appearance. He predicted that, in the summer of 1671, Saturn's ring would be inclined in such a way that it would be viewed edge-on from Earth, and would thus disappear from view. His prediction was correct, confirming his ring theory.

## How were **Saturn's rings formed**?

We are still not sure how Saturn's rings were formed. One idea is that the rings were once larger moons that were destroyed, either by collisions, or by tidal interactions with Saturn's gravity tearing them apart. The bits of moons then settled into orbit around Saturn.

## What are the **physical properties** of **Uranus**?

Uranus is the seventh major planet in our solar system, and the third of four gas giant planets. It is 31,800 miles (51,200 kilometers) in diameter, just under four times the diameter of Earth. Like the other gas giant planets, Uranus consists mostly of gas. Its pale blue-green, cloudy atmosphere is made of 83 percent hydrogen, 15 percent helium, and small amounts of methane and other gases. Uranus gets its color because the methane in the atmosphere absorbs reddish light and reflects bluish-greenish light. Deep down below its atmosphere, a slushy mixture of ice, ammonia, and methane is thought to surround a rocky core.

Although it orbits the Sun in a perfectly ordinary, near-circular ellipse every 84 Earth years, Uranus has an extremely odd rotation compared to the other major planets. It rotates on its side, almost like a bowling ball rolling down its lane, and its polar axis is parallel rather than perpendicular to its orbital plane. This means that one end of Uranus faces the Sun for an entire half of its orbit, while the other end faces away during that time. So one "day" on Uranus is equal to 42 Earth years! Most astronomers think that at some point in its history, Uranus was struck by a large (at least planet-sized) object that knocked it onto its "side," causing this unusual motion.

Uranus is orbited by 15 known moons and 11 thin rings. During its flyby of Uranus, the *Voyager 2* space probe discovered a large and unusually shaped mag-

## What is unique about how was Neptune discovered?

Neptune is the first planet whose existence was first predicted mathematically and then observed afterward. Soon after William Herschel discovered Uranus in 1781, astronomers measured a strange anomaly in its orbit, almost as if a massive object even more distant that Uranus itself were occasionally pulling on the planet. The German mathematician Karl Friedrich Gauss (1777–1855) made calculations based on these planetary movements that laid the groundwork for the discovery of another, more distant planet. In 1843, a self-taught astronomer named John Couch Adams (1819–1892) began a series of complicated calculations that pinpointed the location of such a planet; he finished the calculation in 1845. In 1846, a French astronomer named Urbain Jean Joseph Leverrier (1811–1877) also made a determination of this planet's location. The calculations of Adams matched those of Leverrier, though neither knew of the other's work at the time. On September 23, 1846, Johann Galle (1812–1910) and Heinrich d'Arrest (1822–1875) at the Urania Observatory in Berlin, Germany, found this planet based on the calculations of Leverrier, confirming the findings of both men.

netic field around Uranus (probably unique because of the planet's odd rotational motion) and a chilly cloud-top temperature of –350 degrees Fahrenheit (–210 degrees Celsius).

### Who **discovered Uranus**, and what did he contribute to our understanding of the universe?

The German-born astronomer William Herschel (1738–1822), who lived and worked in England most of his life, was an avid stargazer since his youth. Herschel was conducting a general survey of the stars and planets when, in 1781, he observed a disk-shaped object in the direction of the constellation Gemini. At first Herschel thought the object was a comet. But over time, he observed that its orbit was not elongated as a comet's normally is, but was rather circular, like that of a planet. He wanted to name this new planet George, after King George III of England, but that name did not stick. Eventually, astronomers agreed upon the name Uranus, the mythological father of the Roman god Saturn. In 1787, Herschel also discovered the two largest moons of Uranus.

### What are the **rings of Uranus** like?

The first nine rings of Uranus were discovered in 1977. When *Voyager 2* flew by Uranus in 1986, it found two new rings, bringing the total to 11, plus a number of ring fragments. All are composed of small pieces of dust, rocky particles, and ice. The 11 rings occupy the region between 24,000 and 32,000 miles (38,000 and 51,000 kilometers) from the planet's center. Each ring is between 1 to 1,500 miles

Neptune's distinctive blue color is due to a combination of helium, hydrogern, and methane in its atmosphere. (*NASA*)

(1 and 2,500 kilometers) wide. The presence of ring fragments suggest that the rings may be much younger than the planet they encircle; it is possible that the rings are made of fragments of a broken moon.

The outermost ring, called the epsilon ring, is particularly interesting; it is very narrow and comprised of ice boulders. Two of the small moons of Uranus, Cordelia and Ophelia, act as shepherd satellites to the epsilon ring. They orbit the planet within that ring, and are probably responsible for creating the gravitational field that confines the boulders into the pattern of rings.

## What are the **physical properties** of **Neptune**?

Neptune is the eighth major planet in our solar system, 17 times more massive than Earth and about four times its diameter. The most remote of the four gas giant planets in our solar system, Neptune takes 165 Earth years to orbit the Sun once. A "day" on Neptune, however, is only 16 Earth hours. Similar to Uranus, Neptune's cloud-top temperature is a frosty –350 degrees Fahrenheit (–210 degrees Celsius).

Neptune is bluish-green in color, which might seem fitting for a planet named after the Roman god of the sea. However, the color does not come from water; it is due to the gases in Neptune's atmosphere reflecting sunlight back into space. Neptune's atmosphere consists mostly of hydrogen, helium, and methane. Below the atmosphere, scientists think there is a thick layer of ionized water, ammonia, and methane ice, and deeper yet is a rocky core many times the mass of Earth.

Neptune is so distant that very little was known about it until 1989, when the *Voyager 2* spacecraft flew by Neptune and obtained spectacular data about this mysterious gas giant. Today, we know of at least four ringlets and 11 moons that orbit Neptune.

## What is **Neptune's atmosphere** like?

Despite its distance from the Sun, conditions on Neptune are remarkably active and energetic, which is not what you might expect from a bitterly frigid environment. Neptune is subject to some of the fiercest winds in the solar system, up to 700 miles (1,100 kilometers) per hour. Its layer of blue surface clouds whip around with the wind, while an upper layer of wispy white clouds—probably comprised of methane crystals—rotate with the planet. A darker cloud layer, probably composed of hydrogen sulfide, lies below the methane. The *Voyager 2* flyby of Neptune showed three notable storm systems on the planet: the Great Dark Spot, which is about the size of Earth; the Small Dark Spot, about the size of our Moon; and a small, fast-moving, whitish storm called "Scooter"

> ### How did Phobos and Deimos become Martian moons?
>
> The physical appearances of Phobos and Deimos are very similar to small asteroids. That, and the proximity of Mars to the asteroid main belt, suggests that they were indeed once asteroids whose orbits took them close to Mars. The orbital conditions were just right for Mars to capture them with its gravity, causing them to enter into stable orbits around Mars.

that seems to chase the other storms around the planet. In 1994, however, observations by the Hubble Space Telescope showed that the Great Dark Spot had disappeared.

## What are **Neptune's rings** like?

The flyby of *Voyager 2* past Neptune in 1989 revealed four very faint rings that are less pronounced than those of Saturn or even those of Jupiter and Uranus. These rings are composed of mostly dust particles of varying sizes. The particles in the outermost ring clump together in three places, creating relatively bright, curved segments at three different spots on that ring. That is unlike any other planetary ring in the solar system, and it is not known why this has happened.

# MOONS

## What is a **moon**?

A moon is a natural satellite that orbits a planet. As with planets, it is sometimes hard to know exactly what status a moon has. For example, whereas many moons (such as Earth's Moon) formed at about the same time as the planets they orbit, many other moons probably formed as independent objects that were then captured in a planet's gravitational field.

## How many **moons** does **Mars** have?

Mars has two moons, Phobos and Deimos. They were discovered by the American astronomer Asaph Hall (1829–1907) in 1877.

## What are **Phobos and Deimos** like?

Phobos and Deimos are irregularly shaped rocky objects. They look very much like asteroids. Phobos is about 10 miles across, and Deimos is about half that size.

## What are some of the **characteristics** of **Jupiter's moons**?

Most of Jupiter's dozens of moons (more than 30, as of 2008) are just a few miles across, and are probably captured asteroids. Four of Jupiter's moons stand out,

Jupiter's largest moons were discovered by Galileo and are thus called the Galilean moons. They include (from left to right) Io, Ganymede, Europa, and Callisto. (*NASA*)

however. They are called the Galilean moons because Galileo first discovered them in 1609. The aptly named *Galileo* spacecraft gave humans our closest look yet at these four remarkable planetary bodies, which are richly complex worlds unto themselves.

### What is Jupiter's moon **Io** like?

Io, the closest of the Galilean moons to Jupiter, is affected so strongly by the gravitational tides exerted on it by Jupiter and the other moons that it is the most geologically active body in our solar system. The *Voyager* spacecraft first detected huge volcanoes spewing lava and ash into space, and the surface is completely recoated with fresh lava every few decades.

### What is Jupiter's moon **Europa** like?

Europa is the second closest to Jupiter of the four Galilean moons. Its surface is covered with frozen water ice. Studies by the *Galileo* spacecraft show that the ice has been moving and shifting much the same way that densely packed ice behaves on Earth's polar oceans.

## How does Jupiter influence the conditions on its moons?

Jupiter's tremendous gravitational influence on its surroundings causes tidal activity on the Galilean moons. The tides alternately stretch and compress the cores of these moons, the way you might stretch and compress a soft rubber ball by repeatedly squeezing it in your hand. After a while, the ball will get warmer in your hand from all the physical deformation. The same is true on a planetary scale between Jupiter's gravity and the cores of the Galilean moons.

Another important influence exerted by Jupiter on its moons comes from the giant planet's magnetic field. Jupiter spins so fast, and contains so much mass, that the magnetic field generated by it engulfs the nearby moons and bathes them with ionization and charged particles. Meanwhile, powerful volcanoes that erupt on the surface of Io eject large amounts of small particles into space; many of them are swept up into Jupiter's magnetosphere, forming a doughnut-shaped torus of volcanic particles that form an ethereal envelope around the Jovian environment. (This structure is called, appropriately, the Io torus.)

## What is Jupiter's moon **Ganymede** like?

Ganymede is the largest moon in the solar system, about one-and-a-half times as wide as Earth's Moon. It has a very thin atmosphere and its own magnetic field. Measurements taken by the *Galileo* spacecraft showed atomic hydrogen gas escaping from Ganymede's surface; and other measurements made using the Hubble Space Telescope showed excess oxygen on the surface of Ganymede's thick icy crust. Scientists think that the hydrogen and oxygen may come from molecules of frozen water ice on Ganymede's surface, which are then broken up into their component atoms by radiation from the Sun. These and other observations suggest that, like Io, Ganymede may also have a vast underground sea of water.

## What is Jupiter's moon **Callisto** like?

Callisto, the furthest away from Jupiter of the four Galilean moons, is scarred and pitted by ancient craters. Its surface may be the oldest of all the solid bodies in the solar system. There is evidence here, too (albeit weaker than that in Europa and Ganymede), that a magnetic field may exist around Callisto, which could be caused by a salty liquid ocean far below its surface.

Jupiter's moon Io has many active volcanoes on its surface. (*NASA*)

## What are some of the **characteristics** of **Saturn's moons**?

Like Jupiter, Saturn has dozens of moons. Also like Jupiter, many of these are small moons that are likely to be asteroids captured in Saturn's gravitational field. The larger ones, however, have fascinating characteristics. Mimas, the victim of a huge cratering collision long ago, looks almost exactly like the fictional "Death Star" space station from the movies. Enceladus was recently detected as having geysers of water shooting out from its surface, suggesting the presence of liquid water deep in its core. The most complex moon of Saturn, however—perhaps the most complex moon in the entire solar system—is Saturn's largest moon, Titan.

## What is Saturn's moon **Titan** like?

Titan was discovered by Christian Huygens (1625–1695) around 1655. Over the centuries, astronomers discovered that this largest of Saturn's moons is the only moon in the solar system with a substantial atmosphere—it is even denser than the atmosphere of planet Earth. Titan's atmosphere appears to be composed mainly of nitrogen and methane, with many other ingredients as well. Observations with the space probe *Voyager 1,* and with other telescopes, suggested that Titan might harbor liquid nitrogen or methane at its surface, perhaps in lakes and seas, and that its clouds may produce chemical rains and other weather patterns. Any detailed view is blocked by Titan's thick, opaque atmosphere, however.

The *Cassini* spacecraft launched the *Huygens* probe into the atmosphere of Titan in January 2005. Despite its numbing cold (–300 degrees Fahrenheit), there are topological features that look like tall mountains, rocky beaches, rivers, lakes, and even seas and shorelines. Liquid appears in abundance on the surface of Titan, but it is not liquid water. At those temperatures, water is frozen solid and as hard as granite. Rather, the liquid is probably methane—liquid natural gas.

## What are some of the **characteristics** of **Uranus's moons**?

The moons of Uranus are smallish structures made of ice and rock, ranging in size from about 15 miles (25 kilometers) to 1,000 miles (1,600 kilometers) across. The two largest, Oberon and Titania, were discovered by William Herschel (1738–1822); the next largest two moons, Umbriel and Ariel, were discovered in 1851 by William Lassel (1799–1880). It was not until 1948 that Gerard Kuiper (1905–1973) detected Miranda, the fifth Uranian moon. The *Voyager 2* flew by Uranus in January and February 1986, and discovered at least 10 new moons—all smaller than about 90 miles (145 kilometers) across.

Like the larger moons of Saturn and Jupiter, the five larger moons have varying amounts of geologic features, including craters, cliffs, and canyons. Oberon, for instance, shows an ancient, heavily cratered surface, which indicates there has been little geologic activity there; the craters remain as they were originally formed, and no lava has filled them in. In contrast, Titania is punctuated by huge canyons and fault lines, indicating that its crust has shifted significantly over time.

## What is unique about the moon Triton?

Triton is extremely interesting in that, even though it is the coldest known place in the solar system at –390 degrees Fahrenheit (–235 degrees Celsius), it has a very active environment. It harbors volcanic activity, with several volcanoes shooting not ash, but frozen nitrogen crystals as high as 6 miles (10 kilometers) above the surface. Such eruptions can create temporary layers of haze and clouds over Triton. Scientists think that volcanoes on Triton once covered the moon's surface with a slushy ammonia-and-water-ice "lava," which is now frozen in patterns of ridges and valleys. Triton is also the only major moon in the solar system that orbits in a direction opposite to that of its planet. Triton makes an orbit around Neptune about once every six days. It is possible that Triton was once a large comet-like object, like Pluto, and was captured into Neptune's gravitational field.

## What are some of **Neptune's major moons**?

Triton is Neptune's largest moon; it was discovered soon after Neptune itself was found. The second Neptunian moon to be discovered was Nereid, and that did not happen until 1949. It was discovered by the Dutch-American astronomer Gerard Kuiper. During its 1989 flyby of Neptune, the *Voyager 2* found six other moons, ranging in size from 3 miles (50 kilometers) to 250 miles (400 kilometers) across. At least three more have been discovered since then, all of them very small.

## What are **Pluto's moons** like?

The largest moon of Pluto, Charon, is several hundred miles across. Pluto and Charon are tidally locked to one another, with the same sides always facing one another as they orbit. The other two moons were discovered in 2005, and their existence was confirmed in 2006. Each one is only about 10 miles across.

## What are some of the **largest moons** in the solar system?

The following table lists large moons in our solar system.

### Large Moons in Our Solar System

| Name | Planet | Distance from planet (km) | diameter (km) | orbital period (days) |
|------|--------|--------------------------|---------------|----------------------|
| Moon | Earth | 384,000 | 3,476 | 27.32 |
| Phobos | Mars | 9,270 | 28 | 0.32 |
| Deimos | Mars | 23,460 | 8 | 1.26 |
| Amalthea | Jupiter | 181,300 | 262 | 0.50 |
| Io | Jupiter | 421,600 | 3,629 | 1.77 |
| Europa | Jupiter | 670,900 | 3,126 | 3.55 |

| Name | Planet | Distance from planet (km) | diameter (km) | orbital period (days) |
|------|--------|---------------------------|---------------|-----------------------|
| Ganymede | Jupiter | 1,070,000 | 5,276 | 7.16 |
| Callisto | Jupiter | 1,883,000 | 4,800 | 16.69 |
| Mimas | Saturn | 185,520 | 398 | 0.94 |
| Enceladus | Saturn | 238,020 | 498 | 1.37 |
| Thetis | Saturn | 294,660 | 1,060 | 1.89 |
| Rhea | Saturn | 527,040 | 1,528 | 4.52 |
| Dione | Saturn | 377,400 | 1,120 | 2.74 |
| Titan | Saturn | 1,221,850 | 5,150 | 15.95 |
| Hyperion | Saturn | 1,481,000 | 360 | 21.28 |
| Iapetus | Saturn | 3,561,300 | 1,436 | 79.32 |
| Miranda | Uranus | 129,780 | 472 | 1.41 |
| Ariel | Uranus | 191,240 | 1,160 | 2.52 |
| Umbriel | Uranus | 265,970 | 1,190 | 4.14 |
| Titania | Uranus | 435,840 | 1,580 | 8.71 |
| Oberon | Uranus | 582,600 | 1,526 | 13.46 |
| Proteus | Neptune | 117,600 | 420 | 1.12 |
| Triton | Neptune | 354,800 | 2,705 | 5.88 |
| Nereid | Neptune | 5,513,400 | 340 | 360.16 |

# THE KUIPER BELT AND BEYOND

## What is the **Kuiper Belt**?

The Kuiper Belt (also called the Kuiper-Edgeworth Belt) is a doughnut-shaped region that extends between about three to eight billion miles (5 to 12 billion kilometers) out from the Sun (its inner edge is about at the orbit of Neptune, while its outer edge is about twice that diameter).

## What are **Kuiper Belt Objects**?

Kuiper Belt Objects (KBOs) are, as their name implies, objects that originate from or orbit in the Kuiper Belt. Only one KBO was known for more than 60 years: Pluto. Many KBOs have been discovered since 1992, however, and the current estimate is that there are millions, if not billions, of KBOs.

KBOs are basically comets without tails: icy dirtballs that have collected together over billions of years. If they get large enough—such as Pluto did—they evolve as other massive planetlike bodies do, forming dense cores that have a different physical composition than the mantle or crust above it. Most short-period comets—those with relatively short orbital times of a few years to a few centuries—are thought to originate from the Kuiper Belt.

## Why was a search for Pluto ever initiated?

**A**fter Neptune was discovered in 1846, astronomers measuring its orbit thought they had discovered a strange anomaly, the same sort of measurement found in the orbit of Uranus decades before that led to the discovery of Neptune. In the first decade of the 1900s, the American astronomer Percival Lowell (1855–1916) started to use his observatory near Flagstaff, Arizona, to search for a mysterious "Planet X" that might be causing this orbital anomaly. Lowell became notorious for his idea that the channels observed on Mars may have been a network of canals designed by intelligent living creatures; unfortunately, he did not live to see the discovery of Pluto. However, the Lowell Observatory, which he founded in 1894 in Flagstaff, still exists, and contributes significantly to astronomical research and education to this day.

## What are **Plutinos**?

Plutinos are Kuiper Belt Objects that are smaller than Pluto, have many physical characteristics similar to Pluto, and orbit around the Sun in much the same way that Pluto does. The discovery of Plutinos led to the recognition that the Kuiper Belt is heavily populated, and that Pluto itself is a Kuiper Belt Object.

## What are some of the characteristics of **Pluto**?

Like the other Kuiper Belt Objects, the dwarf planet Pluto is so far away and so small that it is still mysterious in many ways. We do know, though, that Pluto is about 1,400 miles (2,300 kilometers) across, less than one-fifth the diameter of Earth and smaller than the seven largest moons in the solar system. Pluto is composed mostly of ice and rock, with a surface temperature between –350 and –380 degrees Fahrenheit (–210 to –230 degrees Celsius); the bright areas observed on Pluto are most likely solid nitrogen, methane, and carbon dioxide. The dark spots may hold hydrocarbon compounds made by the chemical splitting and freezing of methane.

Pluto's "day" is about six Earth days long, and its "year" is 248 Earth years long. Pluto travels in a highly elliptical orbit around the Sun compared to the terrestrial planets and gas giants. For 20 years out of its 248-Earth-year orbital period, it is actually closer to the Sun than Neptune. (This phenomenon last occurred between 1979 and 1999.) When Pluto is closer to the Sun, its thin atmosphere exists in a gaseous state, and is comprised primarily of nitrogen, carbon monoxide, and methane. For most of its very distant orbit, though, there is no standing atmosphere because it all freezes out and drops to the surface.

Pluto has no rings and three known moons. (Yes, dwarf planets—and even asteroids—can have moons.) The largest one, Charon, is large enough to be considered a dwarf planet in its own right.

## Who **discovered Pluto**?

The American astronomer Clyde Tombaugh (1906–1997), who humbly described himself as a "farm boy amateur astronomer without a university education," was working at the Lowell Observatory when the search for a suspected "Planet X" was started. Tombaugh's job was to continue the demanding task of photographing the area of the sky where this planet was believed to exist. Tombaugh became a celebrity for his momentous discovery of Pluto in 1930, winning a college scholarship for his work. He went on to a distinguished career as an astronomer.

## What is **Eris** and why is it important?

In 2005, astronomers Mike Brown (1965–), Chadwick Trujillo (1973–), and David Rabinowitz (1960–) used a sophisticated, modern version of Clyde Tombaugh's technique to discover a new solar system body beyond the orbit of Pluto and larger than Pluto. Originally called 2003UB 313, this object settled once and for all the question of whether or not Pluto was the largest Kuiper Belt Object in the solar system: it was not. Further observations showed that 2003UB 313 even had its own moon. For a while, this new KBO and its moon were jokingly referred to as "Xena" and "Gabrielle" by its discoverers, in reference to a mythical television heroine and her companion.

The discovery of 2003UB 313 hastened the need for planetary astronomers to define the term "planet" in a scientific way. Since it was larger and more distant than Pluto, Eris would have to be called the tenth planet, unless Pluto was not to be considered a planet any longer. After substantial debate, the objects were officially reclassified in August 2006 by the International Astronomical Union (IAU). That is why there are only eight planets in our solar system today, and why Pluto is not one of them.

Sedna
800-1100 miles
in diameter

Quaoar
(800 miles)

Pluto
(1400 miles)

Moon
(2100 miles)

Earth
(8000 miles)

A comparison of some of the largest Kuiper Belt Objects, including Pluto, Sedna, and Quaoar, compared with Earth and the Moon. (*NASA*)

Not long after this decision, the IAU committee that determines the official names of solar system objects approved the official names of 2003UB 313 and its moon, names that were requested by its discoverers. Today, they are officially known as Eris and Dysnomia, the goddesses of disagreement and argument.

## What are the largest **Kuiper Belt Objects** and how big are they?

The following table lists the largest KBOs in our solar system that are known of today.

### Largest Kuiper Belt Objects

| Name | Geometric Mean Diameter(km) |
| --- | --- |
| Eris | 2,600 |
| Pluto | 2,390 |
| Sedna | 1,500 |
| Quaoar | 1,260 |
| Charon | 1,210 |
| Orcus | 940 |
| Varuna | 890 |
| Ixion | 820 |
| Chaos | 560 |
| Huya | 500 |

# ASTEROIDS

### What is an **asteroid**?

Asteroids are relatively small, primarily rocky or metallic chunks of matter that orbit the Sun. They are like planets, but much smaller; the largest asteroid, Ceres, is only about 580 miles (930 kilometers) across, and only ten asteroids larger than 155 miles (250 kilometers) across are known to exist in the solar system. While most asteroids are made mostly of carbon-rich rock, some are made at least partially of iron and nickel. Aside from the largest ones, asteroids tend to be irregular in shape, rotating and tumbling as they move through the solar system.

### What is the **asteroid belt**?

The asteroid belt (or the "main belt") is the region between the orbit of Mars and the orbit of Jupiter—about 150 to 500 million miles (240 to 800 million kilometers) away from the Sun. The vast majority of known asteroids orbit in this belt. The main belt itself is divided into thinner belts, separated by object-free zones called Kirkwood Gaps. The gaps are named after the American astronomer Daniel Kirkwood (1814–1895), who first discovered them.

### What are the **largest asteroids** in the **asteroid belt**?

The four largest asteroids are the dwarf planet Ceres, Pallas, Vesta, and Hygiea. Other well-known asteroids include Eros, Gaspra, Ida, and Dactyl. The following table lists other asteroids, as well.

#### Largest Asteroids in the Solar System

| Name | Geometric Mean Diameter(km) |
| --- | --- |
| Ceres | 950 |
| Vesta | 530 |
| Pallas | 530 |
| Hygiea | 410 |
| Davida | 330 |
| Interamnia | 320 |
| 52 Europa | 300 |
| Sylvia | 290 |
| Hektor | 270 |
| Euphrosyne | 260 |
| Eunomia | 260 |
| Cybele | 240 |
| Juno | 240 |
| Psyche | 230 |

### How **far apart** are the **asteroids** in the main **asteroid belt**?

Even though there are at least a million or more asteroids in the main belt, the typical distance between asteroids is huge—thousands or even millions of miles. That

## Are all asteroids located in the asteroid belt?

**N**o. There are many asteroids in other regions of the solar system. Chiron, for example, which was discovered in 1977, orbits between Saturn and Uranus. Another example is the Trojan asteroids that follow the orbit of Jupiter near Lagrange points—one group preceding the planet, the other following it—and can thus orbit safely without crashing into Jupiter itself.

means that space chases through the belt, dodging a hail of asteroids, are dramatic but, alas, completely fictional.

## What are **near-Earth objects** and are they **dangerous**?

There are hundreds, if not thousands, of NEOs—near-Earth objects, which are asteroids with orbits that cross Earth's orbit. An NEO could indeed strike our planet, possibly unleashing cosmic destruction.

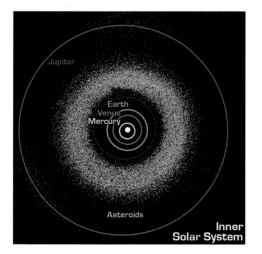

The asteroid belt is located between the orbits of Mars and Jupiter. (*NASA/JPL-Caltech/R. Hurt*)

## When did **astronomers realize** what **asteroids** were?

The first asteroids to be discovered were Ceres (1801), Pallas (1802), Juno (1804), and Vesta (1807). A few decades later, when telescope technology blossomed, the number of tiny, planet-like objects that were found orbiting between Mars and Jupiter mushroomed to dozens, then hundreds. By the middle of the nineteenth century, astronomers realized that these were "minor" planets.

## Where do **asteroids come from**?

The origin of asteroids remains the subject of scientific study. Astronomers today think that most asteroids are planetesimals that never quite combined with other bodies to form planets. Some asteroids, on the other hand, may be the shattered remains of planets or protoplanets that suffered huge collisions and broke into pieces.

## How many asteroids are there?

Today, many thousands of asteroids are being tracked regularly, and tens of thousands have been identified and catalogued. At least one million asteroids are estimated to exist; of those, astronomers estimate that about one in ten can be observed from Earth.

**C**eres was discovered by the Italian priest Giuseppe Piazzi (1746–1826) on January 1, 1801. Piazzi observed a starlike body that was not listed in the star catalogues of the time. He observed the object over several nights and noted that it moved relative to the fixed stellar background, faster than Jupiter but slower than Mars. Piazzi deduced that this object was a new planet, orbiting between Mars and Jupiter. He named the planet Ceres, after the Roman goddess of agriculture. The German mathematician Karl Friedrich Gauss (1777–1855) confirmed Ceres's orbit later that year.

Ceres was considered a planet for several decades, until so many small planets were found orbiting between Mars and Jupiter that astronomers felt that further classification was necessary. Hence, Ceres went from being the smallest planet to the first asteroid ever discovered. Ceres is still the largest asteroid known. Recently, at the same time Pluto's status was adjusted, Ceres's was also adjusted; now, it is considered the largest asteroid and a dwarf planet, as well.

# COMETS

## What is a **comet**?

Comets are basically "snowy dirtballs" or "dirty snowballs"—clumpy collections of rocky material, dust, and frozen water, methane, and ammonia that move through the solar system in long, highly elliptical orbits around the Sun. When they are far away from the Sun, comets are simple, solid bodies; but when they get closer to the Sun, they warm up, causing the ice in the comets' outer surface to vaporize. This creates a cloudy "coma" that forms around the solid part of the comet, called the "nucleus." The loosened comet vapor forms long "tails" that can grow to millions of miles in length.

## When were the **first comets observed**?

Comets can be seen with the naked eye, and sometimes they are spectacularly bright and beautiful, so humans have undoubtedly been observing comets since time immemorial. On the other hand, comets are usually visible only for short periods of time—a few days or weeks—so almost all comet sightings have gone unrecorded and were misunderstood for most of that human history. For these and other reasons, a great deal of mythology and superstition has been associated with comets throughout the ages.

## When did **astronomers calculate** how **comets orbit** the Sun?

By the 1600s, astronomers had reasoned that comets occur in space, beyond Earth's atmosphere, and were trying to figure out where a comet's journey begins and ends.

Johannes Kepler (1571–1630), who observed a comet in 1607, concluded that comets follow straight lines, coming from an infinite distance and leaving forever once they passed Earth. Somewhat later, the Polish astronomer Johannes Hevelius (1611–1687) suggested that comets followed slightly curved paths. In the late 1600s, George Samuel Doerffel (1643–1688) suggested that comets followed a parabolic course. In 1695, Edmund Halley (1656–1742) finally deduced correctly that comets follow highly elliptical orbits around the Sun.

## What was the **first comet** to get a **permanent name**?

The English astronomer Edmund Halley (1656–1742), an acquaintance of Isaac Newton's, was one of the great astronomers of his time. Over his lifetime, Halley created a remarkable legacy of astronomical achievement, even developing the first weather map and one of the first scientific calculations of the age of Earth. Halley served as England's Astronomer Royal, the highest scientific honor in the kingdom at the time, from 1719 to 1742.

One of Halley's greatest discoveries came when he calculated the paths traveled by 24 comets recorded by astronomers over the years. Among these, he found that three—one visible in 1531, one in 1607, and one that Halley himself had observed in 1682—had nearly identical flight paths across the sky. This discovery led him to the conclusion that comets follow in an orbit around the sun, and thus can reappear periodically. In 1695 Halley wrote in a letter to Isaac Newton, "I am more and more confirmed that we have seen that comet now three times, since the year 1531." Halley predicted that this same comet would return 76 years after its last sighting, in the year 1758.

Unfortunately, Halley died before he could see that he was, indeed, correct. The comet was named in his honor, and to this day Halley's comet remains the best-known comet in the world. It last passed by Earth in 1986, and will return again in 2062.

## How did **Heinrich Wilhelm Matthaeus Olbers** help establish a way for calculating the **orbits of comets**?

Since comets orbit in such highly elliptical paths, they can be much harder to calculate than planets and most asteroids. In the late 1700s, the French mathematician and scientist Pierre-Simon de Laplace (1749–1827) had laid down a set of equations to make these calculations, but they were cumbersome and difficult. In 1797 the German astronomer and physician Heinrich Wilhelm Matthaeus Olbers (1758–1840) published a new way to calculate cometary orbits that was more accurate and easier to use than Laplace's technique. The method earned Olbers a reputation as one of the leading astronomers of his time.

Olbers, a highly respected physician who was praised for his vaccination campaigns and for heroically treating people during several epidemics of cholera, set up an observatory in the second floor of his house in 1781. He discovered his first comet in 1780, at the age of 22. Over the course of his lifetime, he discovered five

Comet 73P/Schwassman-Wachmann 3 orbits the sun every five-and-a-half years, and in 1995 the comet fragmented into four pieces, three of which are seen in this 2006 image from the Spitzer observatory. (*NASA/JPL-Caltech/W. Reach*)

comets and calculated the orbits of 18 others. He hypothesized, correctly, that the tail of a comet was created from matter leaving the comet's nucleus and swept back by the flow of energy from the Sun. Olbers was also the discoverer of the second and third asteroids ever found, Pallas in March 1802 and Vesta in March 1807, respectively.

### From **where** do **comets originate**?

Most of the comets that orbit the Sun originate in the Kuiper Belt or the Oort Cloud, two major zones in our solar system beyond the orbit of Neptune. "Short-period comets" usually originate in the Kuiper Belt. Some comets and comet-like objects, however, have even smaller orbits; they may have once come from the Kuiper Belt and Oort Cloud, but have had their orbital paths altered by gravitational interactions with Jupiter and the other planets.

### What is the **Oort Cloud**?

The Oort Cloud is a spherical region enveloping the Sun where most comets with orbital periods exceeding several hundred years (i.e. "long-period comets") originate. The dimensions of the Oort Cloud have never been measured, but it is estimated to be up to a trillion or more miles across. Scientists think that billions, perhaps even trillions, of comets and comet-like bodies are located in the Oort Cloud. Sedna may be the first Oort Cloud object ever discovered.

## Who was the best-known French comet hunter of the eighteenth century?

Charles Messier (1730–1817), the author of the famed Messier catalog, was the best-known French comet hunter of the eighteenth century. His first job was as a draftsman for another astronomer, Joseph Nicolas Delisle (1688–1768), who taught Messier how to operate astronomical instruments. Messier went on to clerk at the Marine Observatory in Paris, and then worked in the tower observatory at the Hotel de Cluny in Paris. From that post, he discovered at least 15 comets and recorded numerous eclipses, transits, and sunspots. Admitted to the French Royal Academy in 1770, he soon produced the first section of his famous catalog of night-sky objects. Among the objects he found, such as the Crab Nebula, he found many comets, as well.

## Who is **Jan Hendrick Oort**?

The Oort Cloud is named for Jan Hendrick Oort (1900–1992), who is widely considered to have been the leading Dutch astronomer of his generation. His scientific research covered a great range of subjects, from the structure of galaxies to the way comets are formed. He was also a pioneer of radio astronomy.

In 1927 Oort investigated the then-revolutionary concept that the Milky Way galaxy is rotating about its center. By studying the motion of stars near the Sun, Oort concluded that our solar system was not at the center of the galaxy, as had been previously believed, but somewhere toward the outer edge. Oort then set out to decipher the structure of the Milky Way, using theoretical models and the tools of radio astronomy.

Oort's work on the origin of comets led him to propose, in 1950, that a huge, shell-shaped zone of space, well beyond the orbit of Pluto and stretching out trillions of miles beyond the Sun in all directions, contains trillions of slowly orbiting, inactive comets. Those comets would remain there until a passing gas cloud or star disturbs the orbit of a comet, sending it toward the Sun and inner solar system in a highly elliptical orbit. Today, this zone of long-period comets bears Jan Oort's name.

## What are some of the **best-known comets** of modern times?

Halley's comet is probably the best-known comet in human history. It last flew by Earth in 1986. Other well-known comets in recent times include Comet Shoemaker-Levy 9, which broke apart and crashed into Jupiter in 1994; Comet Hyakutake, which flew by Earth in 1996; and Comet Hale-Bopp, considered by many to be the "comet of the twentieth century," which flew by Earth in 1997.

## What are some **characteristics** of **Halley's comet**?

Scientists think that Halley's comet is similar to all other comets, except perhaps that it is larger and closer to the Sun than most. In 1986 the European Space

Agency's probe *Giotto* took pictures and other data of the center of Halley's comet. The images showed that the comet was about nine miles (15 kilometers) long and six miles (10 kilometers) wide, coal-black, and potato-shaped, marked by topological features that look like hills and valleys. Two bright jets of gas and dust, each about eight miles (14 kilometers) long, shot out of the comet. The surface of the comet, and the gas in its coma and tail, contained water, carbon, nitrogen, and sulfur molecules.

## What was the **flyby** of **Comet Hale-Bopp** like?

Comet Hale-Bopp was discovered by two astronomers on the same night, which explains the hyphenated name. On July 22, 1995, Alan Hale (1958–) saw the comet from his home in southern New Mexico, and Thomas Bopp (1949–) saw it from Arizona. Comet Hale-Bopp first became visible to the unaided eye in August 1996. It appeared at its brightest for nearly two full months in March and April 1997. Like the spectacular Comet Hyakutake, which had flown by Earth the year before, it also had both a bluish ion tail and a yellowish-white dust tail that curved away from the other tail.

## What happened to **Comet Shoemaker-Levy 9**?

The encounter between Comet Shoemaker-Levy 9 and the planet Jupiter was the first collision between solar system bodies ever directly observed by humans. As the comet approached Jupiter in the spring of 1994, it broke up into a long chain of fragments. Astronomers observed with amazement in July 1994 as these fragments crashed, one by one, into the gas giant's thick atmosphere.

# EARTH AND THE MOON

## EARTH

### What is **Earth**?

Earth is the third planet in the solar system, orbiting at a distance of about 93 million miles (150 million kilometers) from the Sun. It is the largest and most massive of the terrestrial planets. Its interior structure consists of a metallic core, which has both a liquid and solid compenent; a thick rocky mantle; and a thin rocky crust.

### How was **Earth** first **measured**?

The study of the size and shape of Earth is called geodesy. People have studied geodesy for millennia. As early as 2,000 years ago, the Greek-Egyptian astronomer and mathematician Eratosthenes used the shadow of the Sun to compute that Earth was a sphere about 25,000 miles in circumference. This was impressively close to the modern value.

This scientific knowledge was lost and rediscovered several times throughout history, as civilizations rose and fell. By the middle of the fifteenth century, for example, most Europeans living away from the ocean thought that Earth was flat, although sailors and scholars were fully aware that Earth was a sphere. However, the size was still uncertain; Christopher Columbus, for example, thought Earth was much smaller than it actually is. He was convinced he could get to India by traveling west from Spain faster than by traveling east. (He, of course, ran into the Caribbean islands and the Americas instead.)

Finally in the seventeenth and eighteenth centuries, Europeans were able to develop techniques to measure the size and shape of Earth accurately. Dutch physicist, astronomer, and mathematician Willebrord Snell (1580–1626), who is best remembered today for Snell's law, explaining the angle of refraction (bending) of light through different materials, extended these mathematical ideas to figure out how to measure distances using trigonometry. He used a large quandrant (a circular arc divided into 90-degree angles) to measure angles of separation between two points. From this he could calculate distances between them and measure the radius of Earth.

The German mathematician and scientist Karl Friedrich Gauss (1777–1855) also worked on this problem; as director of the Goettingen Observatory from 1807 until the end of his life, Gauss became interested in geodesy. In 1821 he invented the heliotrope, an instrument that reflects sunlight over great distances to mark positions accurately while surveying.

# ORBIT AND ROTATION

### How does **Earth spin**?

Earth's spin is mostly the result of angular momentum left over during the formation process of our planet. There are three distinct motions, the most noticeable being Earth's rotation. Earth rotates once every 23 hours, 56 minutes, causing our cycles of day and night. Earth also has precession (a wobble of the rotational axis) and nutation (a back-and-forth wiggle of Earth's axis), caused primarily by the gravitational pull of the Moon as it orbits Earth. Precession and nutation, over long periods of time, cause Earth's north and south poles to point toward different stars.

### How did **scientists prove** that Earth rotates?

James Bradley (1693–1762), England's Astronomer Royal from 1742 to 1762, first provided evidence of Earth's orbit and rotation. When Bradley tried to measure the parallax of stars—the observed angular motion of a star due to Earth's motion around the Sun—he noticed that all of the stars in the night sky shifted their posi-

## Who was Foucault and how did he come up with the idea for his pendulum?

Jean-Bernard-León Foucault (1819–1868) was a leading scientific figure of his time. Aside from his famous pendulum, Foucault also invented the gyroscope, made the most accurate measurement of the speed of light up to that time, and instituted improvements in the design of telescopes. In addition, Foucault was a prolific writer, producing textbooks on arithmetic, geometry, and chemistry, as well as a science column for a newspaper.

Together with physicist Armand Fizeau (1819–1896), Foucault was the first person to use a camera to photograph the Sun. The camera they used was a daguerreotype, which took pictures on a light-sensitive, silver-coated glass plate. These early plates were barely sensitive to light, compared to the film or digital detectors being used today, so to take their photos, Fizeau and Foucault had to leave the camera focused on Earth for quite a while. It took so long that the Sun's position relative to Earth would change considerably, and the pictures would be blurry. This problem inspired Foucault to invent a pendulum-driven device to keep the camera in line with the Sun.

tions by exactly the same amount throughout the year, and in the same direction that Earth moved. In 1728, it became clear to Bradley that the apparent movement of the stars he observed was due to Earth's forward motion toward the starlight as it came toward Earth. This effect, called the aberration of starlight, is similar to the sensation that makes it seem like raindrops are falling slightly toward an observer as he or she walks through a rainstorm, causing one to tilt an umbrella forward. It showed clearly that Earth was moving, suggesting strongly that Earth was rotating, too.

This famous photo of Earth was taken by astronauts aboard *Apollo 17. (NASA)*

In 1852, the French scientist Jean-Bernard-León Foucault (1819–1868) confirmed Earth's rotation by hanging a large iron ball on a 200-foot (60-meter) wire from the domed ceiling of the Pantheon monument in Paris. A small pointer at the bottom of the ball scratched the ball's path into a flat layer of sand. Over the course of an entire day, the path of the ball remained constant as it swung under the Pantheon; but the line etched out by the pointer slowly and continually shifted to the right. Eventually, the line came full circle, showing a full loop that corresponded with half the length of the day. Foucault's pendulum was a simple, Earth-bound way of proving that Earth's rotation is real, and not an optical illusion caused by the Sun and stars revolving around it.

### How **fast** is **Earth rotating**?

Earth spins around completely once every 23 hours and 56 minutes. This, of course, is not exactly 24 hours; but it is so close that we have created clocks and calendars to reflect the nice round number of 24 hours per day, and compensate in other ways for the difference.

Since Earth is a mostly solid object, every part of Earth takes the same amount of time to complete one rotation. That means, for example, that a person standing on Earth's equator is actually moving in that rotational motion at some 1,040 miles (1,670 kilometers) per hour—nearly twice as fast as a commercial jet liner! This speed goes down as one moves toward the north and south poles, however; at the poles, the rotation speed would be zero.

# THE ATMOSPHERE

### How **thick** is **Earth's atmosphere**?

Earth's atmosphere extends hundreds of miles beyond its surface, but it is much denser at the surface than at high altitudes. About half of the gas in Earth's atmosphere is within a few kilometers of the surface, and 95 percent of the gas is found within 12 miles (19 kilometers) of the surface.

Earth's life-giving atmosphere includes not only the oxygen and carbon dioxide essential for plants and animals, but also upper layers containing ozone and other gases that protect life from harmful radiation. (*NASA*)

### What **gases** form **Earth's atmosphere**?

Earth's atmosphere consists of 78 percent nitrogen, 21 percent oxygen, one percent argon, and less than one percent of other gases, such as water vapor and carbon dioxide.

### What are the various **layers** of **Earth's atmosphere**?

The bottom layer of Earth's atmosphere is called the troposphere. This level is the air we breathe; it contains clouds and weather patterns. Above the troposphere is the stratosphere, which starts at an altitude of about 9 miles (14 kilometers). The temperature in the strato-

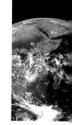

sphere is a frosty –58 degrees Fahrenheit (–50 degrees Celsius). From an altitude of 50 miles (80 kilometers) up to about 200 miles (320 kilometers), temperatures increase dramatically, even though the atmospheric density is very low; this is the thermosphere. Above the thermosphere is the highest layer of Earth's atmosphere: the exosphere or ionosphere. At this level, gas molecules break down into atoms, and many of the atoms become electrically charged, or ionized.

## What is the **mesosphere** and the **ozone layer**?

The mesosphere is the uppermost layer of the stratosphere. Below the mesosphere, at altitudes of 25 to 40 miles (40 to 65 kilometers), is a warm layer of the stratosphere that contains a high concentration of ozone molecules that block ultraviolet light.

## How did **Earth's atmosphere form**?

Some of Earth's atmosphere was probably gas captured from the solar nebula four and a half billion years ago, when our planet was forming. It is thought that most of Earth's atmosphere was trapped beneath Earth's surface, escaping through volcanic eruptions and other crustal cracks and fissures. Water vapor was the most plentiful gas to spew out, and it condensed to form the oceans, lakes, and other surface water. Carbon dioxide was probably the next most plentiful gas, and much of it dissolved in the water or combined chemically with rocks on the surface. Nitrogen came out in smaller amounts, but did not undergo significant condensation or chemical reactions. This is why scientists think it is the most abundant gas in our atmosphere.

The high concentration of oxygen in our atmosphere is very unusual for planets, because oxygen is highly reactive and combines easily with other elements. In order to maintain oxygen in gaseous form, it must constantly be replenished. On Earth, this is accomplished by plants and algae that conduct photosynthesis, removing carbon dioxide from the atmosphere and adding oxygen into it.

# THE MAGNETIC FIELD

## What is **Earth's magnetic field**?

Electromagnetic force permeates our planet. In essence, Earth itself acts like a giant spherical magnet. This is caused primarily by the motion of electrical currents within Earth, probably through the liquid metallic part of Earth's core. Combined with Earth's rotation, the core acts like an electric dynamo, or generator, creating a magnetic field.

Earth's magnetic field extends thousands of miles outward into space. Magnetic field lines, carrying and projecting electromagnetic force, anchor at Earth's magnetic poles (north and south) and bulge outward, usually in large loops. Occasionally, they stream outward into space. The magnetic north and magnetic south poles of Earth's magnetic field are very close to the geographical north and south poles, which mark the axis of Earth's rotation. (Be careful, by the way. There are two ways to define Earth's magnetic poles—the "magnetic north pole" is on an island in Canada, but the "geomagnetic north pole" is actually on Greenland, and the "geographic north pole" is on an ice shelf floating on the ocean, hundreds of miles from any land.)

## How did people **discover** that Earth has a **magnetic field**?

The ancient Chinese were the first to use magnets as compasses for navigation. Though they did not know it, these "south-pointing needles" worked because the magnets aligned themselves with Earth's magnetic field. Since Earth's magnetic poles have been very close to the rotational north and south poles, compasses point almost exactly north and south in most parts of the world.

Over time, scientists started making a connection between lodestones (permanent magnets) and the nature of Earth itself. The English astronomer Edmund Halley (1656–1742), for example, spent two years crossing the Atlantic on a Royal Navy ship, studying Earth's magnetic field. Later, the German mathematician and scientist Karl Friedrich Gauss (1777–1855) made important discoveries about how magnets and magnetic fields work in general. He also created the first specialized observatory for the study of Earth's magnetic field. With his colleague Wilhelm Weber (1804–1891), who was also famous for his work with electricity, Gauss calculated the location of Earth's magnetic poles. (Today, a unit of magnetic field strength is called a gauss in his honor.)

## How **strong** is Earth's **magnetic field**?

On typical human scales, it is pretty weak; at Earth's surface, it is about one gauss in most places. (A refrigerator magnet is typically 10 to 100 gauss.) However, the energy in a magnetic field depends strongly on its volume; so since the field is bigger than our entire planet, overall, Earth's magnetic power is formidable.

## Does Earth's **magnetic field** ever **change**?

Yes, the magnetic field is constantly changing, though very slowly. The magnetic poles actually drift several kilometers each year, often in seemingly random direc-

In 1906 French physicist Bernard Brunhes (1867–1910) found rocks with magnetic fields oriented opposite to that of Earth's magnetic field. He proposed that those rocks had been laid down at a time when Earth's magnetic field was oriented opposite to the way it is today. Brunhes's idea received support from the research of Japanese geophysicist Motonori Matuyama (1884–1958), who in 1929 studied ancient rocks and determined that Earth's magnetic field had flipped its orientation a number of times over the history of our planet. Today, studies of both rock and the fossilized microorganisms imbedded in the rock show that at least nine reversals of Earth's magnetic field orientation have occurred over the past 3.6 million years.

The exact cause of the polarity reversal of Earth's magnetic field is still unknown. Current hypotheses suggest that the reversal is caused by Earth's internal processes, rather than external influences like solar activity.

tions. Over thousands of years, the strength of the magnetic field can go up and down significantly. Even more amazing, Earth's magnetic field can reverse directions—the north magnetic pole becomes the south magnetic pole, and vice versa. According to scientific measurements, our planet's magnetic field last had a polarity reversal about 800,000 years ago.

## What **will happen** when Earth's **magnetic field flips** upside down?

Probably not much will happen to our daily lives when Earth's magnetic field undergoes a polarity reversal. Measurements over the years show that there has been about a six percent reduction in the strength of Earth's magnetic field in the past century, so some scientists think that a polarity reversal on Earth will likely happen sooner rather than later. Some non-scientific hypotheses have been put forth, suggesting that there will be an environmental catastrophe as a result. There is no scientific reason to believe, however, that such disasters will occur.

## Do any **other objects** in the solar system have **magnetic fields** that **flip upside down**?

Yes, all planets and stars with magnetospheres are thought to undergo magnetic polarity reversals. The Sun, for example, undergoes a magnetic field polarity reversal every eleven years. Astronomers can see and study this effect in other astronomical bodies, and from them, learn more about the changes in Earth's own magnetic field.

## What is an **aurora**?

An aurora is a bright, colorful display of light in the night sky. Aurorae are produced when charged particles from the Sun (usually solar wind particles, but sometimes

Solar winds striking Earth's upper atmosphere create the colorful northern and southern lights. (*iStock*)

coronal mass ejections as well) enter Earth's atmosphere. The particles are guided to the north and south magnetic poles by Earth's magnetic field. Along the way, these particles ionize some of the gas molecules they encounter by drawing away electrons from those molecules. When the ionized gas and their electrons recombine, they glow in distinctive colors; and the glowing gas undulates across the sky.

### Where are aurorae seen?

Aurorae, known also as the northern lights (aurora borealis) and southern lights (aurora australis), are most prominent at high altitudes near the north and south poles. They can also be seen sometimes at lower latitudes on clear nights, far from city lights; every once in a while—perhaps once every year or so—aurorae can be seen as far south as the United States. Displays of aurorae can be amazingly beautiful, varying in color from whitish-green to deep red and taking on forms like streamers, arcs, curtains, and shells.

### Do other planets have aurorae?

Every planet with a magnetic field will have aurorae. Beautiful aurorae—sometimes with features larger than the entire planet Earth—have been detected and photographed near the magnetic poles of Jupiter and Saturn.

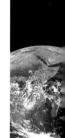

At the Lewis Research Center in Cleveland, Ohio, a scientist at the Electric Propulsion Laboratory creates a simulation generating artificial Van Allen Belts using a plasma thruster. (*NASA*)

# VAN ALLEN BELTS

## What are the **Van Allen belts**?

The Van Allen belts are two rings of electrically charged particles that encircle our planet. The belts are shaped like fat doughnuts, widest above Earth's equator and curving downward toward Earth's surface near the polar regions. These charged particles usually come toward Earth from outer space—often from the Sun—and are trapped within these two regions of Earth's magnetosphere.

Since the particles are charged, they spiral around and along the magnetosphere's magnetic field lines. The lines lead away from Earth's equator, and the particles shuffle back and forth between the two magnetic poles. The closer belt is about 2,000 miles (3,000 kilometers) from Earth's surface, and the farther belt is about 10,000 miles (15,000 kilometers) away.

## How were the **Van Allen belts discovered**?

In 1958 the United States launched its first satellite, *Explorer 1,* into orbit. Among the scientific instruments aboard *Explorer 1* was a radiation detector designed by James Van Allen (1914–2006), a professor of physics at the University of Iowa. It was this detector that first discovered the two belt-shaped regions of the magnetosphere filled with highly charged particles. These regions were subsequently named the Van Allen belts.

## Do **other objects** in the solar system have **Van Allen belts**?

Yes. All the gas giant planets are thought to have such belts, and in Jupiter's magnetic field such belts have been observationally confirmed.

# NEUTRINOS

## What is a **neutrino**?

A neutrino is a tiny subatomic particle that is far smaller than an atomic nucleus; it has no electrical charge and a tiny mass. (Electrons are many thousands of times more massive than neutrinos, and protons and neutrons are many millions of times more massive.) Neutrinos are so tiny and ghostly that they almost always pass through any substance in the universe without any interference or reaction.

## How was the **existence** of **neutrinos proven**?

The existence of neutrinos was first suggested in 1930 by the Austrian physicist Wolfgang Pauli (1900–1958). He noticed that in a type of radioactive process called beta decay, the range of the total energy given off in observations was greater than theorectial predictions. He reasoned that there must be another type of particle present to account for, and carry away, some of this energy. Since the amounts of energy were so tiny, the hypothetical particle must be very tiny as well and have no electric charge. A few years later, the Italian physicist Enrico Fermi (1901–1954) coined the name "neutrino" for this enigmatic particle. The existence of neutrinos was not experimentally confirmed, however, until 1956, when American physicists Clyde L. Cowan, Jr. (1919–1974) and Frederick Reines (1918–1998) detected neutrinos at a special nuclear facility in Savannah River, South Carolina.

## What was the "solar neutrino problem?"

From the very beginning of neutrino astronomy research, there was a discrepancy between the theory of nuclear fusion and the number of neutrinos detected from the Sun. Neutrino telescopes on Earth detected only about half as many neutrinos as they should have. This strange result was checked again and again and repeatedly confirmed. This became known as the solar neutrino problem. Was the Sun generating less energy at its core than expected? Was nuclear fusion theory wrong?

The problem was finally solved nearly four decades after it was first discovered. Neutrinos, as it turns out, can actually change their characteristics when they strike Earth's atmosphere. That meant that there were the right number of neutrinos leaving the Sun, but so many of them changed "flavor" upon reaching Earth that they escaped detection by the neutrino telescopes deep underground. This discovery was a major breakthrough in fundamental physics. It confirmed very important properties about neutrinos that have major implications on the basic nature of matter in the universe.

### If **neutrinos** are so **elusive**, how do **scientists observe** them striking Earth?

It is possible to detect neutrinos from space by their very rare interactions with matter here on Earth, but not with conventional telescopes. The first effective neutrino detector was set up in 1967 deep underground in the Homestake Gold Mine near Lead, South Dakota. There, the American scientists Ray Davis, Jr. (1914–2006) and John Bahcall (1934–2005) set up a tank filled with 100,000 gallons of nearly pure perchlorate (used as dry-cleaning fluid), and monitored the liquid for very rare neutrino interaction events. Other experiments have since used other substances, such as pure water, for neutrino detections.

### Where are the **neutrinos coming** from?

The vast majority of neutrinos striking our planet come from the Sun. The nuclear reactions at the core of the Sun create huge numbers of neutrinos; and unlike the light that is produced, which takes thousands of years to flow their way out of the Sun's interior, the neutrinos come out of the Sun in less than three seconds, reaching Earth in just eight minutes.

### Have **neutrinos** ever been shown to have hit Earth **from somewhere other than the Sun**?

In 1987, the first supernova visible to the unaided eye to occur in centuries appeared in the southern sky. At almost exactly that same moment, neutrino detectors around the world recorded a total of nineteen more neutrino reactions than usual. This worldwide detection does not sound like much, but it was hugely significant because it was the first time neutrinos were confirmed to have reached Earth from a specific celestial object other than the Sun.

# COSMIC RAYS

## What are **cosmic rays**?

Cosmic rays are invisible, high-energy particles that constantly bombard Earth from all directions. Most cosmic rays are protons moving at extremely high speeds, but they can be atomic nuclei of any known element. They enter Earth's atmosphere at velocities of 90 percent the speed of light or more.

## Who **first discovered cosmic rays**?

The Austrian-American astronomer Victor Franz Hess (1883–1964) became interested in a mysterious radiation that scientists had found in the ground and in Earth's atmosphere. This radiation could change the electric charge on an electroscope—a device used to detect electromagnetic activity—even when placed in a sealed container. Hess thought that the radiation was coming from underground and that at high altitudes it would no longer be detectable. To test this idea, in 1912 Hess took a series of high-altitude, hot-air balloon flights with an electroscope aboard. He made ten trips at night, and one during a solar eclipse, just to be sure the Sun was not the source of the radiation. To his surprise, Hess found that the higher he went, the stronger the radiation became. This discovery led Hess to conclude that this radiation was coming from outer space. For his work on understanding cosmic rays, Hess received the Nobel Prize in physics in 1936.

## How were **cosmic rays** shown to be **charged particles**?

In 1925 American physicist Robert A. Millikan (1868–1953) lowered an electroscope deep into a lake and detected the same kind of powerful radiation that Victor Franz Hess had found in his balloon experiments. He was the first to call this radiation cosmic rays, but he did not know what they were made of. In 1932, the American physicist Arthur Holly Compton (1892–1962) measured cosmic-ray radiation at many points on Earth's surface and found that it was more intense at higher latitudes (toward the north and south poles) than at lower latitudes (toward the equator). He concluded that Earth's magnetic field was affecting the cosmic rays, deflecting them away from the equator and toward Earth's magnetic field. Since electromagnetism was now shown to affect the rays, it was clear that cosmic rays had to be electrically charged particles.

## **Where** do **cosmic rays** come **from**?

A continuous stream of electrically charged particles flows from the Sun; this flow is called the solar wind. It makes sense that some fraction of cosmic rays originate from the Sun, but the Sun alone cannot account for the total flux of cosmic rays onto Earth's surface. The source for the rest of these cosmic rays remains mysterious. Distant supernova explosions could account for some of them; another possibility is that many cosmic rays are charged particles that have been accelerated to enormous speeds by interstellar magnetic fields.

# METEORS AND METEORITES

## What is a **meteorite**?

A meteorite is a large particle from outer space that lands on Earth. They range in size from a grain of sand on up. About 30,000 meteorites have been recovered in recorded history; about 600 of them are made primarily of metal, and the rest are made primarily of rock.

## What is a **meteor**?

A meteor is a particle from outer space that enters Earth's atmosphere, but does not land on Earth. Instead, the particle burns up in the atmosphere, leaving a short-lived, glowing trail that traces part of its path through the sky. Like meteorites, meteors can range from the size of a grain of sand on up; most of the time, though, a meteor larger than about the size of a baseball will reach Earth, in which case we call it a meteorite.

## **Where** do **meteors and meteorites** come **from**?

Most meteors, especially those that fall during meteor showers, are the tiny remnants of comets left in Earth's orbital path over many, many years. Most meteorites, which are generally larger than meteors, are pieces of asteroids and comets that somehow came apart from their parent bodies—perhaps from a collision with another body—and orbited in the solar system until they collided with Earth.

## What is a **meteor shower**?

An artist's concept of a meteor burning up as it enters Earth's atmosphere. (*iStock*)

Meteors are often called "shooting stars" because they are bright for a moment and move quickly across the sky. Usually, a shooting star appears in the sky about once an hour or so. Sometimes, though, a large number of meteors appear in the sky over the course of a several nights. These meteors will appear to come from the same part of the sky, and dozens or hundreds (sometimes even thousands) of meteors can be seen every hour. We call such dazzling displays meteor showers. The strongest meteor showers are sometimes called meteor storms.

## How did **scientists learn** that **meteors and meteorites** come from **outer space**?

In 1714 English astronomer Edmund Halley carefully reviewed reports of meteor sightings. From the reports, he calculated the height and speed of the meteors, and deduced that they must have come from outer space. Other scientists, however, were hesitant to believe this notion, thinking instead that meteors and meteorites were either atmospheric occurrences like rain, or debris spewed into the air by exploding volcanoes.

In 1790 a group of stony objects showered part of France. Georg Christoph Lichtenberg (1742–1799), a German physicist, assigned his assistant Ernst Florens Friedrich Chladni (1756–1827) to investigate the event. Chladni examined reports of these falling stones, as well as records over the previous two centuries. He, like Edmund Halley, also concluded that the chunks of matter came from outside Earth's atmosphere. Chladni guessed that meteorites were the remains of a disintegrated planet.

In 1803 in a series of loud explosions, more than 2,000 meteorites fell to Earth onto French territory. Jean-Baptist Biot (1774–1862), a member of the French Academy of Science, collected reports from witnesses as well as some of these fallen stones. Biot measured the area covered by the debris and analyzed the composition of the stones, showing that they could not have originated in Earth's atmosphere.

## How **old** are **meteorites**?

Most meteorites are billions of years old and have been orbiting in the solar system for a very long time before they collide with Earth. Many meteorites are as old as the solar system itself, about 4.6 billion years, and have been largely unchanged for that whole time.

## Where are **meteorites found**?

Meteorites have been found pretty much all over the world. With modern civilization likely to have disturbed most landing sites, meteorites are most likely found today in remote, barren areas like deserts. The majority of meteorites have been discovered in the largest uninhabited, undisturbed part of the world left—Antarctica.

## What **kinds** of **meteorites** are there?

There are two main categoreies of meteorites: stony and metallic. Each category is further subdivided into more detailed groups with similar characteristics. The Vestoids, for example, are all thought to have come from the asteroid Vesta, where, long ago, a powerful collision created shattered bits of Vesta that have been orbiting the solar system ever since. Chondrites are one kind of stony meteorite; they are often the oldest meteorites. Another category, the pallasites, have fascinating mixtures of stony and metallic material. Pallasites probably originated from boundary areas in larger asteroids, where rocky mantles were in physical contact with metallic cores.

## What can **scientists learn** from **meteorites**?

Since meteorites are so old, scientists study them to learn about the early history of our solar system in much the same way that paleontologists study fossils to learn about life on Earth millions of years ago. Some of the oldest stony meteorites even contain grains of material that are older than the solar system.

Metallic meteorites can also be used to learn about the insides of planets like our own. One kind of meteorite, for example, contains both metal and minerals locked

Barringer Crater in Arizona is one of only a few craters on Earth that have not yet been eroded away. Sometimes simply called Meteor Crater, it offers clear evidence of a meteorite impact. (*iStock*)

together in beautiful and complex patterns. Scientists study these objects, called pallasites, to gain insights on the internal structure of Earth near its metallic core.

### Are **falling meteors** and **meteorites dangerous**?

Typical meteors and meteorites pose no danger of any kind to people. Meteors burn up before they reach Earth, so they do not hit anything on the surface; meteorites are so rare that the chances of their hitting anything important are almost zero.

Still, occasional incidents are known to happen. A falling meteorite killed a dog in Egypt in 1911; another struck the arm of—and rudely awakened—a sleeping woman in Alabama in 1954; and in 1992 a meteorite put a hole through a Chevy Malibu automobile. Once in a very rare while—every 100,000 years or so—a meteor or meteorite about 100 meters across will collide with Earth. Once in a very, very rare while—every 100 million years or so—a meteorite 1000 meters across will do so, and that is indeed dangerous.

### What is the **largest known meteorite** to **strike Earth** in the past **100,000 years**?

About 50 thousand years ago, a metallic meteorite about 100 feet across crashed into the Mogollon Rim area in modern-day Arizona. It disintegrated on impact, creating a hole in the desert nearly a mile across and nearly 60 stories deep. Meteor Crater (or the Barringer Meteor Crater, as it is more commonly known today) is a remarkable and lasting example of the amount of kinetic energy carried by celestial objects. Just the lip of the crater rises 15 stories up above the desert floor. For a long

time, scientists puzzled over the origin of this crater. It might have been volcanic in origin, they thought. But geological evidence, such as shallow metallic remnants in a huge radius miles around the crater, confirmed it was a meteorite strike.

## What is the **largest known meteorite** to **strike Earth** in the past **100 million years**?

About 65 million years ago, a meteorite about 10 kilometers (6 miles) across crashed into our planet near what is now southern Mexico. The remanant of this collision is an underwater crater more than 100 miles across. This asteroid, or comet, carried ten million times more kinetic energy than either the Tunguska or Meteor Crater impactors. The heat from the explosion probably set the air itself on fire for miles around. It threw so much of Earth's crust into the atmosphere that it blocked most of the Sun's light for months. As it fell back through the atmosphere, this debris grew very hot as it landed; it probably set almost every tree, bush, and blade of grass it touched on fire. The ecological catastrophe caused by this titanic meteorite strike was most likely the evolutionary blow that finished off the dinosaurs.

# THE MOON

## What is the **Moon**?

The Moon is Earth's only natural satellite. It is 2,160 miles (3,476 kilometers) across, which is a little more than one quarter of Earth's diameter, or about the distance from Cleveland, Ohio, to San Francisco, California. The Moon orbits Earth once every 27.3 days.

The Moon has no atmosphere and no liquid water at its surface, so it has no wind or weather at all. On the lunar surface, there is no protection from the Sun's rays, and no ability to retain heat like the greenhouse effect on Earth. Temperatures on the moon range from about 253 degrees Fahrenheit (123 degrees Celsius) to –387 degrees Fahrenheit (–233 degrees Celsius). The Moon's surface is covered with rocks, mountains, craters, and vast low plains called maria ("seas").

## What is the **Moon made of**?

Though the full Moon sometimes looks very much like a wheel of bleu or Stilton cheese, it is actually covered with rocks, boulders, craters, and a layer of charcoal-colored soil. The charcoal-colored soil consists primarily of pulverized rocky and glassy fragments, and is up to several meters deep. Two main types of rock have been found on the moon: basalt, which is hardened lava, and breccia, which is soil and rock fragments that have melted together. Elements found in moon rocks include aluminum, calcium, iron, magnesium, titanium, potassium, and phosphorus. Unlike iron-rich Earth, the Moon appears not to have much metallic content.

## How **far away** is the **Moon**?

On average, the Moon is about 238,000 miles (384,000 kilometers) away from Earth. This value was measured quite accurately by the ancient Greek astronomer Hipparchus, who lived in the second century B.C.E. Today, laser rangefinders have been used to measure a very precise value.

## How was our **Moon formed**?

The formation of the Moon was a great scientific mystery for many years. It was once thought that Earth and the Moon might have formed simultaneously as two separate objects, bound together by their mutual gravitational pull. This was shown to be unlikely after scientists proved that the two objects have very different compositions. Another idea suggested that Earth's Moon formed elsewhere, and was later captured into Earth's orbit as it went by Earth's gravitational influence. The major problem with this scenario is that Earth and the Moon are relatively close in size; gravitational capture is very, very unlikely, unless one object is many times larger than the other.

Within the past few decades, scientists have shown that the most likely scenario of how the Moon formed involves the collision of two planetary bodies. Billions of

years ago, before life formed on Earth, a Mars-sized protoplanet slammed into Earth at an angle. Most of the material in the protoplanet fell into, and became part of, our planet; some material, however, was thrown out into space, and began to orbit Earth as a ring of dust and rock. Within weeks, a large portion of that ring of material coalesced to form the core of our Moon; over millions of years, the Moon settled into its present-day size and shape.

### How did the **Moon evolve** after it was formed?

A view of the Moon's surface taken from *Apollo 10* shows a crater-pocked, barren landscape. (*NASA*)

Scientists think that, for about the first billion years or so after the Moon formed, it was struck by great numbers of meteorites, which blasted out craters of all sizes. The energy of so many meteorite collisions caused the Moon's crust to melt. Eventually, as the crust cooled, lava from under the surface rose up and filled in the larger cracks and crater basins. These younger regions, which look darker than the older, mountainous areas, are the "seas" (*maria* in Latin) on the Moon.

### Who were the **first astronomers** to **study** the **surface** of our **Moon**?

Galileo Galilei, the first astronomer to study the universe with a telescope, observed that the Moon's surface was not smooth, but rather covered with mountains and craters. The broad, dark patches on the Moon looked to him like seas on Earth, so he named them *maria*, or "seas" in Latin.

### Who were the **first astronomers** to map the **Moon's craters**?

In 1645, Polish astronomer Johannes Hevelius (1611–1687) charted 250 craters and other surface features on the Moon. Today, he is known as the founder of lunar topography. Also around that time, Italian physicist Francesco Maria Grimaldi (1618–1663) built a telescope and used it to make hundreds of drawings of the Moon, which he pieced together to form a map of the features of the Moon's surface.

### How do the **craters** on the **Moon** get their **names**?

Many of the features on the Moon, including many of the large maria, have names from antiquity. Individual craters are typically named after famous people, especially astronomers and other scientists. Official names are approved and recorded by a special committee of the International Astronomical Union.

When seen from Earth, the Moon's landscape appears to some observers to look like a human face; hence the expression "Man on the Moon." (*iStock*)

## Has the **Moon** always been in its **present position** from Earth?

No, the Moon used to be much closer to Earth, and used to go around Earth in a much shorter time than it does today. In the future, the orbital distance between Earth and the Moon will increase; the angular momentum of the Earth-Moon system will dissipate so much so that the Moon will start spiraling toward, and ultimately crash into, our planet. According to calculations, though, long before this occurs our Sun will evolve into a red giant about five billion years from now, destroying the Earth-Moon system.

## What is the **"Man on the Moon"**?

The Man on the Moon is in the eye of the beholder. As viewed from Earth, there appear to be several very large craters and maria (such as Mare Imbrium, Mare Serenitatis, Mare Tranquilitatis—the Seas of Rain, Serenity, and Tranquility) that can look like eyes, a mouth, and other facial features to an imaginative observer.

Over the millennia, different societies and civilizations found their own interpretations of the patterns of craters and maria on the Moon. In southwestern Native American cultures, the Man on the Moon was actually Kokopelli, a large-headed, thin-bodied man hunched over playing the flute. In ancient China, the design of the Moon was not a man at all, but a rabbit.

## What is the **"dark side"** of the **Moon**?

The dark side of the Moon is a misnomer for the side of the Moon that always faces away from Earth. Over billions of years, the rotation of the Moon has become synchronized with its orbit around Earth, so that the same side of the Moon always faces our planet. (That is why the designs people see on the Moon can change orientation, but never change shape.) This phenomenon, called tidal locking, means that the other side of the Moon is never visible from Earth. Though this side is sometimes dark, just as often it is brightly lit by the Sun. So the scientifically accurate way to refer to the Moon's "dark side" should be "the far side" of the Moon.

## Why is the **Moon so bright**?

Moonlight is reflected sunlight. This was discovered long ago by the ancient Greek astronomer Parmenides, who lived and worked around 500 B.C.E. Depending on the location of the Moon in its orbit around Earth, different parts of the Moon will reflect sunlight onto Earth. Since Earth and the Moon are so close together, and

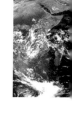

**What other scientist observed the Moon around Galileo's time,
but is not widely credited for his work?**

Interestingly, an Englishman named Thomas Harriot (1560–1621) also
observed the Moon with a telescope a few months before Galileo did. Harriot, who is best known today as a mathematician who made advancements in
the equations and notations of algebra, made his own telescope and observed
Halley's comet, sunspots, and Jupiter's moons. Unlike Galileo, however, Harriot did not record or publish much of his work. Galileo did, and also performed follow-up studies of his discoveries. Thus, Galileo is credited with
being the discoverer of the Moon's craters.

since the Moon has such a shiny surface, large amounts of sunlight come to Earth
after bouncing off the Moon.

## Why does the **Moon** seem to **change shape**?

As viewed from Earth, the amount of sunlight that strikes the Moon and reaches us
changes continually in a periodic (repeating) pattern. That is because Earth orbits
the Sun, whereas the Moon orbits Earth; so the relative positions of the Moon,
Earth, and the Sun keep changing. This kind of regular changing pattern causes the
phases of the Moon.

## How do the **lunar phases work**?

The new moon occurs when the Moon is between Earth and the Sun. All of the sunlight striking the Moon bounces away from Earth, so we do not see any of the Moon
at all. Over the next two weeks or so, the phase of the Moon changes from new to
waxing crescent, then first quarter, then waxing gibbous, until Earth is between the
Sun and the Moon. At that point, all of the sunlight striking the Moon bounces
toward Earth, so we see the entire disk of the Moon. This phase is called the full
moon. Then, over the next two weeks after the Moon is full, the phase changes to
waning gibbous, then third quarter (also called last quarter), then waning crescent,
until the phase of the Moon is new again.

## How long is the **cycle** of the **phases of the Moon**?

The Moon orbits Earth every 27.3 days, while Earth orbits the Sun every 365.25
days. Those facts, combined with the fact that moonlight is actually reflected sunlight, causes the Moon to take on different phases over a 29.5-day cycle.

## How **strong** is the **Moon's gravity** on **people**?

Although the Moon is very massive—73.5 billion billion metric tons—it is so far
away from Earth (240,000 miles or 384,000 kilometers) that it has very little grav-

itational pull on objects at or near Earth's surface. It produces about 1/300,000th the gravitational acceleration that Earth produces at its own surface—far too weak to be felt by any person.

# TIDES

### Does the **Moon's gravity affect Earth** at all?

Definitely! Although the Moon's gravitational pull at any one place on Earth is very weak, the combined effect of the Moon over a large area or volume on Earth can be very noticeable. The Moon's effect is most easily seen in the ocean tides.

### What are **tides**?

Tides are the consequence of any two objects that exert gravitational pull on one another over a long period of time. Basically, each object gently pulls the other object into an egg-like shape, because the gravitational acceleration on one side of the object is larger than on the other side. On Earth, the most observable evidence of this gravitational effect is the changing tides we witness.

### How do **tides work**?

Two cycles of high and low tides occur each day, roughly 13 hours apart. High tides occur both where the water is closest to the Moon, and where it is farthest away. At the points in between, there are low tides.

### How **often** do **ocean tides occur** on Earth?

During a 26-hour period, each point on Earth's surface moves through a series of two high tides and two low tides—first high, then low, then high again, then low again. The length of the cycle is the sum of Earth's period of rotation, or the length of its day (24 hours), and the Moon's eastward orbital movement around Earth (two hours).

### Does the **Sun** also **influence tides** on Earth?

Yes, the Sun also influences Earth's ocean tides, but only about half as much as does the Moon. Although the Sun is many millions of times more massive than the Moon, it is also about 400 times farther away from Earth than the Moon is. Tidal effects, like gravitational force in general, are very sensitive to changes in distance.

## What is a "spring tide"?

When the Moon is in its new phase, or its full phase, the Moon, Earth, and the Sun fall roughly along a straight line in space. As a result, the tidal effects on Earth's oceans are magnified compared to other times in the Earth/Moon/Sun orbit. We call tides that occur at these times the "spring tides," even though they can happen any time of year and have nothing to do with the seasons. The term actually comes from the German *springen,* meaning "to jump" or "to rise up."

## What is a "neap tide"?

When the Moon is in its first quarter or last quarter phase, the line between Earth and the Moon is at right angles to the line between Earth and the Sun. As a result, the tidal interaction between Earth and these two solar system bodies do not work together at all, and the difference between high tide and low tide is the smallest for that month during these times. We call tides that occur at these times "neap tides."

## How does the **tidal action** of the **Moon affect Earth**?

The liquid core of Earth—and, minimally, the solid part as well—is also pulled ever so slightly backward and forward by the Moon's tidal action. Its motion is tiny—much, much less than ocean tides—but over billions of years that kind of tidal activity is like squeezing a rubber ball in your hand over and over; the core heats up. That heat eventually diffuses through the planet, affecting processes like volcanism and plate tectonics.

## How has **Earth's tidal action influenced** the **Moon**?

Earth's tidal action on the Moon actually causes the Moon's spin to slow down. The Moon used to spin on its axis, just like Earth does today, but tidal forces have drained away a large amount of its spin—known in physical terms as "angular momentum." Today, the Moon always has the same side facing Earth.

## What will **ultimately happen** to the **Earth-Moon system** because of their mutual **tidal action**?

If Earth and the Moon were to continue orbiting around one another undisturbed for an indefinite period of time, their mutual tidal action would continue to dissipate their angular momentum. Eventually, Earth will be tidally locked to the Moon, so the same face of Earth will always face the same face of the Moon. Even now, the

Moon's tidal action is slowing down the rate of Earth's spin; a million years from now, a day on Earth will be about 16 seconds longer than it is today.

# CLOCKS AND CALENDARS

## How did the relative **motions of Earth, the Moon, and the Sun** lead to the modern **calendar system**?

The astronomers of the ancient world noticed that three lengths of time were regular and predictable: the cycle of nighttime and daytime (a day), the cycle of the lunar phases (a month), and the cycle of the amount of daylight per day over a large number of days (a year). These ancients did not realize that a day is the time it takes Earth to rotate once about its axis; that a month is the time it takes the Moon to make one orbit around Earth; and that a year is the time it takes Earth to make one orbit around the Sun. Astronomers eventually figured out these relative motions of Earth, the Moon, and the Sun, and further refined their timekeeping. For example, they realized that the difference between the Moon's orbital period (27.3 days) and the cycle of the Moon's phases (29.5 days) was caused by the additional motion of Earth around the Sun.

Eventually, days, months, and years were all subdivided into units based on their utility, and based on long-standing customs and traditions. The differences between the common time units and the astronomical motions that spawned them are made compatible through the use of devices such as leap-years and leap-seconds.

## Who established the length of the **calendar solar year**?

As far back as 5,000 years ago, the ancient Egyptians had already established a calendar with 365 days in a solar year. They divided the year into 12 months of 30 days each, and added five additional days at the end of each year. Millennia later, the Danish astronomer Tycho Brahe (1546–1601) determined the exact length of the solar year to a precision of one part in 30 million—an accuracy of one second per year!

## Who established the length of the **calendar solar day**?

The ancient Egyptians originally based the length of the solar day on nightly observations of a series of 36 stars (called decan stars), which rose and set in the sky at 40 to 60 minute intervals. For 10 days, one particular star would be the first decan to appear in the sky, rising a little later each night until a different decan star would be the first to rise. Thus, the first "hours" were marked nightly by the appearance of each new decan in the sky. Depending on the season, between 12 and 18 decans were visible throughout the course of a night. Eventually, the official designation of the hours came at midsummer, when 12 decans (including Sirius, the Dog Star) were visible. This event coincided with the annual flooding of the Nile River—a crucial event in the ancient Egyptian civilization. Thus, the night was eventually divided into 12 equal parts. The 12 daylight hours were marked by a sundial-like device—a notched, flat stick attached to a crossbar. The crossbar would cast a shadow on successive notches as the day progressed. Eventually, the combination of the 12 hours of the day and 12 hours of the night resulted in the 24-hour day we use today.

## What were the **origins** of our **modern annual calendar**?

The original model for our calendar was created by the ancient Romans and Greeks as far back as the eighth century B.C.E. With the help of the Roman astronomer Sosigenes, Julius Caesar created what is known as the Julian calendar in 46 B.C.E. This was the first calendar with a leap year, and a day was added every fourth year. This meant that each year was 365.25 days long. The Julian calendar was off by only 11 minutes and 14 seconds each year when compared to the actual orbit of Earth around the Sun. That is pretty impressive, but over the centuries it added up until, by the sixteenth century, the calendar was off by nearly eleven days.

## When was the **modern calendar established**?

In 1582 Pope Gregory XIII consulted with astronomers and decreed another change in the calendar to remove the 11 minutes and 14 seconds' difference between the Julian year's length and Earth's orbital period around the Sun (which is very close to 365.2422 days). First, the Gregorian calendar reset the date ahead by 10 days, so that the first day of spring would be March 21 of each year. Then, it reduced the number of leap-year days by three days every four centuries. This was accomplished by modifying the leap-year rule: if a year is divisible by 4, it would be a leap-year, unless the year is also divisible by 100. If a year is divisible by 100, it would only be a leap-year if it is also divisible by 400. That means that the year 1600 and the year 2000 were both leap years, but the years 1700, 1800, and 1900 were not—and the years 2100, 2200, and 2300 will not be leap years either.

The Gregorian calendar is the basis of the modern calendar. It is accurate to within 26 seconds per year on average (0.0003 days). To keep everything on track over the long term, by international agreement every once in a while a leap second is added to the end of a year. These adjustments should keep the calendar on target for many thousands more years.

## How does the **lunar phase cycle** coincide with our **modern calendar system**?

Though most of our daily lives are scheduled according to the solar calendar—that is, based on the motion of Earth around the Sun—the cycle of lunar phases influences our lives a great deal as well. Many holidays with ancient origins were scheduled according to calendars based on lunar phases, so that is how we still decide the dates of such occasions as Easter, Passover, Hanukkah, Ramadan, and the Chinese New Year.

# THE SEASONS

## What is the **ecliptic plane**?

The ecliptic plane is the plane of Earth's orbit around the Sun. Ancient astronomers were able to trace the ecliptic as a line across the sky, even though they did not know Earth actually orbited the Sun. They merely followed the position of the Sun compared to the position of the stars in the sky, figured out (despite the Sun drowning out the light of the other stars) where the Sun was every day, and noticed that every 365 days or so the positions would overlap and start going over the same locations again. That line marked a loop around the celestial sphere. Astronomers marked the line using twelve zodiac constellations positioned near and through the loop.

## What is the **difference** between the **ecliptic plane** and Earth's **equatorial plane**?

The equatorial plane is the plane of Earth's equator extended indefinitely out into space. It turns out that Earth's rotation around its axis is not lined up with the ecliptic plane. Instead, Earth is tilted about 23.5 degrees. This tilt is the main cause of the seasons on Earth.

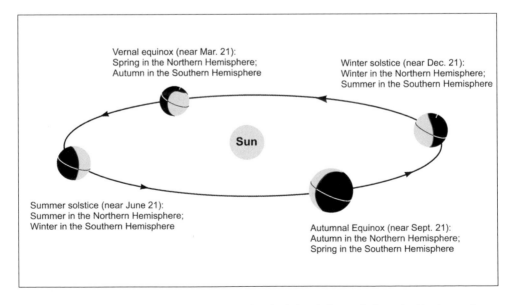

Vernal equinox (near Mar. 21):
Spring in the Northern Hemisphere;
Autumn in the Southern Hemisphere

Winter solstice (near Dec. 21):
Winter in the Northern Hemisphere;
Summer in the Southern Hemisphere

Sun

Summer solstice (near June 21):
Summer in the Northern Hemisphere;
Winter in the Southern Hemisphere

Autumnal Equinox (near Sept. 21):
Autumn in the Northern Hemisphere;
Spring in the Southern Hemisphere

Because Earth is tilted on its axis, either the southern or northern hemisphere is closer to the Sun as it orbits, thus creating the seasons.

## When do the seasons start and end?

**W**hen it comes to climate and weather, the seasons start at different times of year depending on where one is on Earth. Astronomically speaking, though, the first day of spring happens on the vernal equinox; the first day of summer happens on the summer solstice; the first day of fall happens on the autumnal equinox; and the first day of winter happens on the winter solstice.

## How does the **motion** of **Earth around the Sun** cause the **seasons** to occur?

Some people mistakenly think that the seasons are caused by Earth being farther from the Sun in winter and closer to the Sun in summer. This is incorrect; Earth's elliptical orbit is close enough to a perfect circle that distance is not the reason. In fact, Earth is closest to the Sun in early January and farthest in early July, which is exactly the opposite of our summer and winter seasons.

The reason for the seasons has to do with the angle at which sunlight strikes any particular place on Earth at any given time of year. The angle changes throughout the year because the tilt of Earth's axis differs from the ecliptic. Put another way, the equatorial plane and the ecliptic plane are tilted with respect to one another by about 23.5 degrees. When one part of Earth is tilted toward the Sun, that part experiences summer; when it is tilted away from the Sun, it experiences winter; in between these phases Earth experiences spring and autumn.

## What are **solstices** and when do they occur?

A solstice is a time of the year when Earth is pointed either the closest toward the Sun or the farthest away from it. On the summer solstice, there are more minutes of daylight than there are on any other day of the year; on the winter solstice, there are fewer minutes of daylight than there are on any other day of the year. In the northern hemisphere, the summer solstice occurs around June 21 of each year, when the North Pole is pointed closest toward the Sun, and the winter solstice occurs around December 21 of each year, when the North Pole is pointed farthest away from the Sun.

## What are **equinoxes** and when do they happen?

An equinox is a time of the year when, in the course of Earth's orbit, our planet is at a location where the equatorial plane and the ecliptic plane intersect. In other words, the tilt of Earth's axis is pointed perpendicular to the line between Earth and the Sun at an equinox—Earth's poles are tilted neither "toward" nor "away" from the Sun, but tilted off to the "side." On the day of an equinox, there are as many minutes of daylight as there are of night—hence the term "equinox," meaning "equal darkness." In the northern hemisphere, the vernal (spring) equinox occurs around March 21 of each year, and the autumnal (fall) equinox occurs around September 21.

The phases of a lunar eclipse. (*iStock*)

# ECLIPSES

### What is an **eclipse**?

An eclipse is the partial or total blocking of the light of one object by another. In our solar system, the relative positions of the Sun, Moon, and Earth create solar eclipses and lunar eclipses. Total solar eclipses are particularly beautiful.

### How **often** do **eclipses** occur?

Perfect alignments of the Sun, Moon, and Earth are relatively uncommon, because the plane of Earth's orbit around the Sun (called the ecliptic plane) is not the same as the plane of the Moon's orbit around Earth. Thus, during the new moon or full moon phases when an eclipse might be possible, the Moon is usually located just above or below the straight line that runs between Earth and the Sun, so no eclipse occurs. All three objects—Earth, Moon, and Sun—line up just right about twice a year.

### How does a **lunar eclipse** occur?

A lunar eclipse occurs when Earth passes between the Sun and the Moon in such a way that the Moon moves into Earth's shadow. When a partial lunar eclipse is going on, the curved shadow of our planet is apparent on the Moon's face; the Moon looks kind of like it is in a crescent phase, but the terminator line (the line between light and dark) is not curved the same way. When a total lunar eclipse is happening, the entire Moon is in Earth's shadow, and the Moon looks full, but glows only faintly red.

## What is the best way to view a lunar eclipse?

The best way to view a lunar eclipse is, as some astronomers jokingly say, the way you watch paint dry. Lunar eclipses last for hours from start to finish, and no protective equipment is needed.

## How long do **lunar eclipses last**, and how widely is a lunar eclipse visible?

Lunar eclipses tend to last for several hours, from beginning to end. Totality—the time when the Moon is in the darkest part of Earth's shadow, and Earth blocks all direct sunlight onto the Moon—usually lasts for the better part of an hour. Any given lunar eclipse can be seen from everywhere on the planet where it is nighttime.

## Why is the **Moon still visible** during **totality** of a lunar **eclipse**?

Earth's atmosphere is dense enough to act a little bit like a lens, so it refracts a small amount of sunlight shining through it toward the Moon. This small fraction of light, which is mostly red because that is the color of light that refracts best, bounces off the Moon's surface and comes back to Earth. Before and after totality, the direct sunlight reflected off the Moon is so strong by comparison that it drowns out this refracted light, so we normally cannot see it with our unaided eyes. During totality, however, the Earth-atmosphere-refracted light is quite visible as a soft reddish glow.

## What are some upcoming **total lunar eclipses** and where will they be viewable?

The table below lists the next lunar eclipses through the year 2020.

### Total Lunar Eclipses, 2008–2020

| Date | Where Visible |
| --- | --- |
| December 21, 2010 | east Asia, Australia, Pacific, Americas, Europe |
| June 15, 2011 | South America, Europe, Africa, Asia, Australia |
| December 10, 2011 | Europe, east Africa, Asia, Australia, North America |
| April 15, 2014 | Australia, Pacific, Americas |
| October 8, 2014 | Asia, Australia, Pacific, Americas |
| April 4, 2015 | Asia, Australia, Pacific, Americas |
| September 28, 2015 | east Pacific, Americas, Europe, Africa, west Asia |
| January 31, 2018 | Asia, Australia, Pacific, west North America |
| July 27, 2018 | South America, Europe, Africa, Asia, Australia |
| January 21, 2019 | central Pacific, Americas, Europe, Africa |

## How does a **solar eclipse** occur?

A solar eclipse happens when the Moon is directly in line between Earth and the Sun. The Moon's shadow sweeps across Earth's surface; at those places where the shadow lands, an eclipse is seen. Like Earth's shadow, the Moon's shadow consists of two parts: a dark, central region called the umbra, and a lighter region called the penumbra that surrounds the umbra. Under the penumbra, a partial solar eclipse occurs. Under the umbra, a total eclipse or an annular eclipse is seen.

### How **long** do **solar eclipses last**, and how widely is a solar eclipse visible?

The entire process of a solar eclipse, from the beginning of partial coverage until the end, usually takes about an hour. However, the totality of solar eclipse lasts at most only a few minutes. Most total solar eclipses last between 100 and 200 seconds—just about two to three minutes. Furthermore, total solar eclipses can be observed only from narrow bands on Earth's surface, and these bands change with each eclipse. In any given location on Earth, therefore, a total solar eclipse may appear only once every few centuries.

In a total solar eclipse it is possible to get a good look at the Sun's corona, as long as you take measures to protect your eyes from the harmful rays of the Sun. (*iStock*)

### What does a **total solar eclipse look like**?

During totality of a solar eclipse, the Sun looks like a perfectly black disk surrounded by glowing light. This light is actually the Sun's corona, which is invisible under normal circumstances because the Sun is so bright. Away from the corona, the sky is dark, so planets and stars that ordinarily could be seen only at night become visible.

### What is the **best way** to **observe** a **solar eclipse**?

The Sun's light is so powerful that looking at it for too long, even during any part of a partial solar eclipse, can cause

> **Why does the Moon block the Sun so perfectly in a solar eclipse
> that the corona is visible, but the Sun itself is not?**
>
> The Moon's diameter is just under 400 times smaller than the diameter of
> the Sun. Coincidentally, the Moon's distance from Earth is also just under
> 400 times smaller than the Sun's distance from Earth. That is why the Moon
> covers almost exactly the same amount of sky, when viewed from Earth's sur-
> face, as the Sun. This is also why total solar eclipses are so beautiful, with a
> black-as-ink solar disk surrounded by a shimmering, ethereal solar corona.

permanent eye damage. Do not ever look directly at the crescent of a nearly total solar eclipse without proper eye protection. Special sun viewing glasses or filters made of thick mylar or welder's glass can be used, but be absolutely sure those filters are properly rated for viewing the Sun, and that they are not damaged in any way.

One safe, indirect way to look at a partial solar eclipse—or the Sun at any other time, for that matter—is by using a simple pinhole camera. Take two pieces of cardboard, one of which has a white surface. Make a small hole in one card by piercing it with a pin. Turn your back to the Sun and hold the card with the pinhole up so that sunlight enters the hole. Now hold the other card, with the white surface facing up, below the first card so that the image of the sunlight through the pinhole lands on the surface. Adjust the distance between the two cards, and bring the Sun's image into focus. Now you can watch the bottom card to follow the progression of the eclipse behind you.

The one time it is safe to look directly at the Sun without eye protection is during totality of a total solar eclipse. It will only be a few minutes at most, but if you are lucky enough to be there, enjoy the view! Take lots of pictures too, if you have the chance. During totality, ordinary unfiltered cameras will also be unharmed.

## What are some upcoming **total solar eclipses** and where can they be seen?

The following table lists total solar eclipses and their locations through 2020.

### Total Solar Eclipses, 2008–2020

| Date | Where Visible |
| --- | --- |
| July 22, 2009 | India, Nepal, China, central Pacific |
| July 10, 2010 | south Pacific, Easter Island, Chile, Argentina |
| November 13, 2012 | north Australia, south Pacific |
| November 3, 2013 | Atlantic, central Africa |
| March 20, 2015 | north Atlantic, Faeroe Islands, Svalbard |
| March 9, 2016 | Sumatra, Borneo, Sulawesi, Pacific |
| August 21, 2017 | north Pacific, United States, south Atlantic |
| July 2, 2019 | south Pacific, Chile, Argentina |
| December 14, 2020 | south Pacific, Chile, Argentina south |

# SPACE PROGRAMS

## ROCKET HISTORY

### In space exploration, what does **"space" mean**?

In the parlance of space travel, NASA officially defines "outer space" as anywhere beyond an altitude of 62 miles (100 kilometers) above Earth's surface. This is very different from the astronomical definition of "space," which, according to the general theory of relativity, means the curvable three-dimensional fabric within which objects in the universe are situated.

### How are **space vehicles launched** into space?

Rockets are the only way humans have been able to launch objects from Earth into space so far. A rocket is a vehicle system that carries all of its own propellant. The propellant is accelerated to a high speed—usually by combustion, converting it into gas and heating it up—and pushed out the back of the rocket as exhaust. The rocket, following Newton's third law of motion, is pushed forward by the motion of the exhaust.

Most launch vehicles consist of a series of successively smaller rockets placed one on top of the other. The largest rockets provide the most thrust, but are also heaviest; so once their fuel is expended, they are released away from the other, smaller rockets, which then have much less mass to push. With this successive downscaling of the launch vehicle's mass, the payload—usually a spacecraft or satellite—can reach the speeds high enough to get into space, reach orbit, or escape Earth's gravity toward other locations or objects in the universe.

### How have **rockets** developed **before** the **twentieth century**?

Around 160 C.E., the ancient Greek mathematician Hero of Alexandria created a spinning, spherical, steam-powered device that first demonstrated the idea of rocket-like propulsion using hot gas exhaust. Real rockets, though, were first used by the Chinese, who invented the first solid propellant—gunpowder—in the ninth century. In thirteenth-century China, simple hand-held rockets (called "fire arrows") were set off during religious ceremonies and celebrations. These inaccurate, short-range devices were fueled by a mixture of potassium nitrate (saltpeter), charcoal, and sulfur. Their use eventually spread throughout Asia and Europe.

Beginning in the eighteenth century, rockets began to become effective weapons of war. The French military were the leading rocketeers of the time, though mostly for fireworks. Then, in the 1790s, Indian soldiers used rockets to defeat the British army in a number of battles. These rockets weighed about 10 pounds, were attached to sharp bamboo sticks, and could travel about a mile. Although individually these early rockets were very inaccurate, they were intimidating weapons when fired in large barrages at large targets. In 1804, British army officer William Congreve developed rockets that could travel almost two miles. It was these rockets, and the red glare they produced over the city of Baltimore, Maryland, and Fort McHenry, that helped inspire the American poet Francis Scott Key to write *The Star-Spangled Banner,* which a century later became the national anthem of the United States.

### Who **pioneered** the design and construction of **rockets** for **space**?

The Russian engineer Konstantin Tsiolkovsky (1857–1935), American scientist Robert Goddard (1882–1945), and German physicist Hermann Oberth (1894–1989) are generally considered to be the three main visionaries behind the rockets used for modern spaceflight. Though the three men never worked together, their parallel efforts were eventually synthesized to create the international space programs of the twentieth and twenty-first centuries.

### What important **advances** did **Konstantin Tsiolkovsky** propose?

Konstantin Tsiolkovsky (1857–1935) conducted experiments on air travel years before the first powered aircraft was flown by the Wright brothers in 1903. He constructed Russia's first wind tunnel to study airflow acting on airplanes as they flew. In 1895 he introduced his ideas of space travel, and three years later he outlined many of the basic concepts of rocketry and space travel that scientists still use today. A true visionary, Tsiolkovsky was far ahead of any other scientist in this field. He wrote, for example, that humans could survive in space only if supplied with oxygen inside a sealed cabin. In 1903 he published an article titled "The Exploration of Cosmic Space by Means of Reaction Devices," which detailed his ideas about rocket propulsion and the use of liquid fuels.

### What important **advances** did **Robert Goddard** propose?

Robert Goddard (1882–1945) was fascinated by the prospect of space travel and rocketry at an early age. In 1919, he published a now-classic work on his rocketry research, *Method of Reaching Extreme Altitudes,* in which he proposed the possi-

Dr. Robert Goddard (far right and inset) after a 1932 test in Mexico on a rocket including a gyroscope of his design. Goddard held patents on liquid rocket propellants and multi-stage rockets. He anticipated the use of rockets to reach the Moon and developed rockets that reached heights of over one mile high. (*NASA*)

bility that a rocket could eventually reach the Moon. After a number of experiments and early failures, Goddard launched the world's first liquid-fuelled rocket in 1926. This 10-pound (4.5 kilogram) rocket, launched from a cabbage patch in Augurn, Massachusetts, went up 41 feet (12 meters) and traveled a distance of 184 feet (56 meters). Over the next two decades, he greatly advanced the field of rocketry, working out systems for various stages of rocket flight from ignition and fuel system to guidance controls and parachute recovery. In 1930 Goddard set up the world's first professional rocket test site in Roswell, New Mexico, successfully launching rockets as high as 1.3 miles (2.0 kilometers). NASA's Goddard Space Flight Center is named in honor of this rocketry pioneer.

## What important **advances** did **Hermann Oberth** propose?

Hermann Oberth (1894–1989) was born in Transylvania, in what is modern day Romania, and built his first rocket as a teenager. He obtained his higher education in Germany; for his doctoral dissertation, he wrote about the mathematical theories of rocketry and practical considerations of space flight. This work, *By Rocket into Planetary Space,* was rejected by his German academic advisors, but later revised and expanded into the book *Ways to Spaceflight,* which was published in 1929. Over his career, Oberth worked on solid propellant rockets and on vehicles to go to the Moon.

## How are typical **rocket engines configured**?

Rockets are configured to rapidly combine fuel and oxidizer—two ingredients necessary for combustion—so there is tremendous burning or even explosions. The resulting hot, expanding gas must be controlled in its output, both in amount and direction. The key is the exit of exhaust gases from the rocket, which is allowed to escape through ports or nozzles in one end. The forceful exit of the exhaust creates thrust that moves the rocket upward and forward.

## How are **rocket engines fueled**?

Most rockets are fueled by liquid propellant, a mixture of liquid fuel and liquid oxidizer. These two substances are stored in the rocket, but in separate tanks. They are combined in a combustion chamber, where they are ignited and produce the energy that propels the vehicle. Typical liquid rocket fuels include alcohol, kerosene, hydrazine, and liquid hydrogen; typical liquid oxidizers include nitrogen tetroxide or liquid oxygen. Some rockets use solid rather than liquid propellant. In this case, the oxidizer and fuel are already combined in a dormant solid state. When the mixture is ignited, the entire amount of propellant is consumed in a single controlled combustion reaction. Solid-fuel rockets are generally made to have more thrust than liquid-fuel rockets; also, they are lighter, simpler to design, and do not have nearly as many moving parts. Liquid-fuel rockets, on the other hand, can be turned on and off, and the amount of their thrust can be carefully controlled for performing delicate maneuvers.

## How **powerful** are typical space **rockets** today?

Rockets today have varying sizes, masses, and lift capacities, and different ones are used depending on the payload that is being delivered into space. Typical rockets that are used to send supplies or other small payloads to the International Space Station or other low Earth-orbit destinations include the Soyuz-Fregat system, which sits on the launch pad to a height of about 120 feet (35 meters), weighs about 300 tons fully loaded with fuel, and can produce about 800,000 pounds of thrust at takeoff. Systems that launch exploratory spacecraft (like *Messenger, Cassini,* and the Mars Exploration Rovers) include the Delta II rockets, which stand about 120 feet high and deliver about 1,000,000 pounds of thrust at takeoff, and the Atlas V rockets, which are about 190 feet tall and can deliver up to about 2,000,000 pounds of thrust.

### Who **pioneered** the world's **first** successful **space program**?

In the Soviet Union, the man credited with creating the world's first successful space program was the Ukranian scientist Sergei Korolëv (1906–1966). In 1931 Korolëv became director of the rocket research group in Moscow; he worked there for many years, but his work was interrupted by World War II. After the war ended, he returned to rocket research and helped incorporate captured German technology into the Soviet rocket program. His work bore bountiful fruit: in August 1957, he launched the first Russian intercontinental ballistic missile (ICBM). Less than two months later, a rocket based on the ICBM was used to launch *Sputnik 1,* the first man-made satellite to orbit Earth. In 1959 *Luna 3* was the first space probe to send back pictures of the far side of the Moon. Then, in 1961, Korolëv led the design and construction of *Vostok 1,* which carried the first human being into space: Yuri Gagarin (1934–1968); and in 1963, the first woman, Valentina Tereshkova, was launched into space. In 1966, the *Venera 3* mission was the first spacecraft to land on another planet—Venus—and the *Luna 9* probe was the first spacecraft to land on the Moon. Korolëv was so important that the government of the Soviet Union kept his identity secret—referring to him only as "Chief Designer of Launch Vehicles and Spacecraft"—until after his death in 1966. He was buried in the Kremlin Wall, an honor reserved only for the most distinguished Soviet citizens.

### Who **pioneered** the **space program** in the **United States**?

The scientist generally considered to be the most influential figure in the American space program was the German physicist Wernher von Braun (1912–1977). Born into a wealthy family, von Braun became an amateur astronomer at an early age, and studied at the University of Berlin. One of his mentors was the German rocketry pioneer Hermann Oberth. Soon after the Nazis came to power in Germany, von Braun was placed in charge of research and devel-

After working on rockets for the Germans during World War II, Dr. Wernher von Braun was recruited by the Americans to work on the U.S. space program. (*NASA*)

opment of rockets as weapons for the German military. Under his leadership, the Germans developed the V-2 rocket, the first long-range rocket-launched missile weapon system.

Near the end of World War II, von Braun and 126 other German scientists were hired by the United States government and brought to America under the code name Project Paperclip. Using captured German rockets, the scientists taught their American counterparts about their rocketry; they also continued their rocket research and test flights at White Sands Proving Grounds in New Mexico, and at Fort Bliss in Texas. A few years later, they were moved to NASA's Marshall Space Flight Center in Huntsville, Alabama. There, von Braun was named the center's first director and presided over the construction of a new long-range ballistic missile called the Redstone. Eventually, von Braun led the effort to create the *Jupiter-C—* the first American rocket capable of launching spacecraft. This rocket launched America's first satellite into orbit, *Explorer 1.* It was followed by the *Saturn V,* which was used to launch the Apollo manned missons to the Moon.

# SATELLITES AND SPACECRAFT

### What do **satellites** and **spacecraft** need to **operate**?

Once a satellite or spacecraft is launched, it still needs a propulsion system to move or turn; communications and telemetry systems to send and receive data and commands from scientists and flight controllers; and an electrical energy system to power everything on the craft. For every spacecraft, the way these systems are designed and operated differs depending on the payload of the craft.

### How do **satellites** and **spacecraft move** in space?

Once in space, satellites need only a small amount of force to move—either to speed up or slow down, or to change direction and orientation. Small rocket engines

A xenon ion engine being tested at NASA's Jet Propulsion Laboratory in Pasadena, California. (*NASA*)

(often referred to as thrusters) are usually sufficient. Even small rocket thrusters, however, require large amounts of fuel over time, which weighs down the spacecraft and reduces the payload. Spacecraft designers now use new technologies to move things around in space, such as ion propulsion engines.

## How does an **ion propulsion engine** work?

An ion propulsion engine uses magnetic fields rather than chemical combustion to achieve thrust. A small amount of gas—usually of a heavy element such as xenon— is injected into an ionization chamber containing a series of magnetic coils. An electrical power supply powers the coils, and the resulting electromagnetic forces in the chamber separate the positively and negatively charged particles in the gas, creating ions and free electrons. Using powerful electric fields, those charged particles are then accelerated to very high speeds, and then pushed out of the back of the ionization chamber. Their backward motion creates forward thrust.

## How **powerful** are **ion propulsion systems**?

Compared to typical chemical combustion rockets, the force generated by an ion propulsion system is weak. Current ion engines on spacecraft generate less thrust at full power than a child pushing a toy truck with her hand. But ion engines are so efficient that they use very little fuel, even at full power. They can, therefore, last for years and continue pushing for days, weeks, or even months at a time.

## What is the typical **electrical power source** for **spacecraft**?

For most satellites and spacecraft that operate inside the orbit of Mars, solar panels are the easiest way to generate electrical power. They convert sunlight into electric current, which can then be stored in batteries for use by the craft's many power-consuming activities. Beyond a few hundred million miles from the Sun, however, solar panels do not work very well because the sunlight is too weak. For those more distant missions—such as *Galileo, Cassini,* and the *Voyager* spacecraft—a power source called a radioisotope thermoelectric generator (RTG) has historically been very effectively used.

## How does an **RTG work**?

Radioisotope thermoelectric generators (RTGs) are not nuclear power plants in space. They are heavily shielded containers that hold several kilograms of radioactive isotopes such as plutonium-238, and include equipment that converts the heat released by the isotopes' radiative decay into electrical power. The *Cassini* spacecraft, for example, carries three RTG units, each of which started off with about 17 pounds (eight kilograms) of plutonium-238 and generated about 300 watts of electric power.

## Are there **nuclear power plants** on today's **spacecraft**?

No nuclear reactors are known to be aboard any currently operational satellites or spacecraft. Late in the twentieth century, the Soviet Union launched a number of military satellites that contained compact nuclear reactors. A few of them, however, nearly ended in disaster. *Cosmos 954,* which was launched in September 1977, spiraled into the atmosphere and crashed into the Canadian arctic on January 24, 1978, scattering radioactivity across a large stretch of land. Some of the debris was emitting lethal doses of radiation when it was recovered. Fortunately, nobody was killed or injured, but cleanup of the affected area took months. Another Soviet spacecraft launched in August 1982 called *Cosmos 1402,* suffered the same fate, falling to Earth on January 23, 1983. Fortunately, this re-entry occurred far out in the Indian Ocean, and no known debris was ever found.

Today, for safety reasons, no spacecraft are launched with nuclear reactors onboard. It remains a very attractive idea, however, to have nuclear-powered spacecraft for deep space journeys that carry them far from Earth. The challenge is to make sure that, even with a catastrophic failure, our planet and its people would not be put at risk. In the 1960s, the United States studied a nuclear-powered rocket engine idea called NERVA (Nuclear Engine for Rocket Vehicle Application), but the project was cancelled in 1972. In 2003 NASA started a new nuclear space program called Project Prometheus; after a few years, however, its funding was deeply cut, and its future is in doubt.

# THE SPUTNIK ERA

### What was the **first man-made object** to **orbit Earth** in space?

On October 4, 1957, the former Soviet Union launched the first artificial satellite into orbit around Earth. It was called *Sputnik 1,* the Russian word meaning "traveling companion" or "satellite." During its three months in space, *Sputnik 1* orbited the planet once every 96 minutes, at a speed of nearly 17,400 miles (28,000 kilometers) per hour. The Soviets' success caught U.S. engineers—and the American public—by surprise, and launched the so-called "space race" between the two rival world superpowers of that time.

### What was the *Sputnik 1* satellite like?

*Sputnik 1* was a steel ball 23 inches (58 centimeters) in diameter and weighing 184 pounds (83 kilograms). Attached to its surface were four flexible antennae, ranging from 2.2 to 2.6 yards (201 to 238 centimeters) long. *Sputnik 1* transmitted radio signals at two frequencies and gathered valuable information about the ionosphere and temperatures in outer space.

### What was the **first satellite** launched by the **United States**?

The United States had a space program nearly ready to go when *Sputnik 1* was

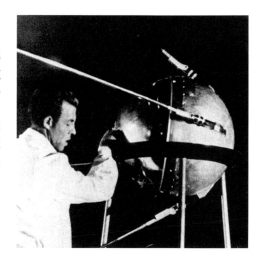

The first artificial satellite successfully launched into space was the Soviet Union's *Sputnik 1,* launched on October 4, 1957. (*Asif A. Siddiqi*)

201

launched in 1957. Startled into action, the American government hurriedly rushed to launch the first orbiter, called *Vanguard,* on December 6, 1957. The launch was a failure: the rocket carrying the satellite burst into flames just a few feet off the ground. The following month, on January 31, 1958, a team led by Wernher von Braun at the Marshall Space Flight Center in Huntsville, Alabama, successfully launched *Explorer 1,* the first American satellite, into orbit on the nose of a Jupiter-C rocket.

### What was the *Explorer 1* satellite like?

*Explorer 1* was a bullet-shaped satellite about 6.5 feet (2 meters) long and weighing 31 pounds (14 kilograms). It was designed by the pioneering space scientist James Van Allen (1914–2006) at the University of Iowa, and contained instruments to measure the temperature and density of Earth's upper atmosphere. It also had a radiation detector that found thick rings of radiation surrounding our planet, which today are called the Van Allen belts. *Explorer 1* remained in orbit until 1967, returning valuable scientific data about the nearby reaches of outer space.

### What happened in the *Sputnik* era of spaceflight?

After *Sputnik* and *Explorer 1,* the Soviet and American space programs continued to launch satellites. The United States' Vanguard program was only somewhat successful, with only three successful launches of spacecraft in 11 attempts. The Explorer program was more successful; a total of 65 Explorer spacecraft were launched between 1958 and 1984, which provided detailed pictures of Earth as viewed from space, as well as a great deal of data on a range of space phenomena, including solar wind, magnetic fields, and ultraviolet radiation. Meanwhile, the Sputnik program continued with four more launches between 1957 and 1960.

# COMMUNICATIONS SATELLITES

### Who conceived the idea of a **communications satellite**?

The idea of using an orbiting satellite for communication was first introduced by the British science fiction writer Arthur C. Clarke (1917–2008). In 1945, he proposed constructing an international communication system using three orbiting satellites. To make this idea a reality, however, scientists had to overcome a number of technical obstacles. The satellites, and the equipment onboard, would have to withstand extreme heat and cold, and have a power supply that could last for years without replacement. Then it was a matter of sending this communication equipment into orbit!

## What were the **first communications satellites**?

*Sputnik 1,* the first artificial satellite ever launched, had communications capabilities. It was able to transmit radio signals at two frequencies. It lasted for about three months in orbit.

The first long-lived communications satellite was called *Echo,* and was launched in 1960. Developed by John R. Pierce (1910–2002) of Bell Telephone Laboratories, *Echo* was an aluminum-coated, gas-filled plastic balloon 100 feet (31 meters) across. It was placed in a low orbit and passively reflected communications signals, bouncing them back to Earth without any active transmission. Its successor, *Echo II,* was in service from 1964 to 1969.

The first active-transmitting communications satellites were *Telstar,* developed by AT&T Corporation, and *Relay,* developed by NASA. *Telstar* was launched in 1962 and transmitted telephone calls and television broadcasts between locations in Maine, England, and France. Together, *Telstar* and *Relay* demonstrated the potential of multi-satellite communications systems for long-distance global transmissions.

## What is **INTELSAT**?

Seeing the need for a comprehensive, jointly owned and operated system of satellite communications, 11 nations formed the International Telecommunications Satellite Organization—INTELSAT—on August 20, 1964. On April 6, 1965, *Early Bird,* the organization's first satellite and the first commercial communications satellite, was launched. It was a metal cylinder a foot and a half tall and two feet wide, and was encircled by a band of solar cells. It could handle 240 telephone lines or one television channel at a time. Over the years, more nations joined the organization, and many more satellites were launched. In 2001 INTELSAT became a private company, Intelsat Limited. Today, it continues to provide satellite communications services with its fleet of more than 50 satellites.

## What is the **Global Positioning System (GPS)**?

Today, there are hundreds of satellites in orbit around Earth—many of them are

Thanks to dozens of satellites orbiting Earth, Global Positioning Systems are possible for use in cars or hand-held devices that keep people from becoming lost. (*iStock*)

communications satellites that transmit phone, audio, television, and other electromagnetic signals all around the globe. One of the best known communications satellite systems is the NAVSTAR Global Positioning System, or GPS. This is a system of 24 satellites orbiting Earth at an altitude of about 12,000 miles (19,300 kilometers) and speeds of 7,000 miles (11,260 kilometers) per hour. By obtaining simultaneous communications signals from several satellites at once, it is possible to pinpoint the location of a GPS receiver to a precision of just a few feet or meters anywhere on Earth. The GPS system is maintained by the U.S. government, at a cost of more than $700 million per year. The economic and social benefit of this system, however, far exceeds this amount.

# FIRST HUMANS IN SPACE

### Who was the **first human** in **space**?

The first person to go to outer space was the Russian cosmonaut Yuri Gagarin (1934–1968). He was attending the Industrial Technical School at Saratov, Russia, on the Volga River, when he joined a flying club and became an amateur pilot. At the recommendation of an instructor, Gagarin was accepted to the Orenbury Aviation School in 1955. On November 7, 1957, Gagarin graduated with honors and was given the rank of lieutenant. He then went off to the Arctic to train as a fighter pilot. Inspired by the successful 1959 flight of the Soviet satellite *Luna 3*, which orbited the Moon, he applied to be among the first group of cosmonauts and was approved. For more than a year, he was involved in the testing and training for spaceflight.

On April 12, 1961, Gagarin was launched in the Soviet spacecraft *Vostok 1*. His journey into space lasted 108 minutes, during which time *Vostok 1* orbited Earth once and then returned. When his capsule was about two miles above the ground, Gagarin parachuted to safety and became a hero to the world. For the five years following his historic flight, Gagarin was kept busy with public appearances, political activities, administrative tasks, and training the next group of cosmonauts. In 1966 he began to prepare himself for another space mission, this time aboard a Soyuz spacecraft. The first Soyuz flight took place the following year; unfortunately, the cosmonaut onboard, Vladimir Komarov (1927–1967), was killed during reentry into Earth's atmosphere. Gagarin continued to train nonetheless, but never got the chance to go into

The first human in space was Soviet hero Yuri Gagarin. (*NASA*)

space again. During a training flight on March 27, 1968, his aircraft spun out of control and crashed, killing him and his flight instructor.

## Who was the **first woman** in **space**?

The Soviet-born Valentina Tereshkova (1937–) was an accomplished amateur parachutist by 1961, when she applied to join the Soviet space program. She was one of the first four female cosmonauts selected into the program, and in 1963 she piloted *Vostok 6* for three days, orbiting Earth 48 times. During the flight, a smiling Tereshkova was shown on Soviet and European television, signaling that all was well. "I see the horizon," she said. "A light blue, a beautiful band."

Valentina Tereshkova returned to a heroine's welcome, and was awarded the title "Hero of the Soviet Union." She toured the world, and eventually attained the rank of colonel in the Soviet Air Force. She also completed a technical science degree, and served as an aerospace engineer in the Soviet space program. She entered politics, and became a high-ranking official in the Soviet government. After the fall of the Soviet Union, Tereshkova chaired the Russian Association of International Cooperation. She married fellow cosmonaut Andrian Nikolayev, and their daughter Elena was born in 1964—the first person ever born to parents who had both been in space!

## Who was the **first American** in **space**?

On May 5, 1961, American astronaut Alan Bartlett Shepard Jr. (1923–1998) made history with the first piloted American space flight. Shepard rode on the Mercury-Redstone 3 mission, in the *Freedom 7* spacecraft, which flew on a sub-orbital trajectory. He reached an altitude of 116 miles (187 kilometers), and traveled a distance of 303 miles (488 kilometers) in space at a speed of 5,146 miles (8,280 kilometers) per hour. Shepard's 15-minute flight ended with a safe parachute landing into the Atlantic Ocean.

Shepard was a navy pilot, and achieved the rank of Rear Admiral. He remained in the astronaut corps after his historic first flight. He eventually commanded the *Apollo 14* mission to the Moon, and was the fifth person to walk on the Moon's surface.

## Who was the **first American** to orbit Earth?

On February 20, 1962, John H. Glenn Jr. (1921–) became the first American to orbit Earth. Glenn's historic flight was part of the Mercury program, which was initiated by NASA to send humans into space. Glenn traveled inside a capsule

Alan Shepard, the first American in space, is hauled aboard a helicopter after successfully splashing down to Earth in his *Freedom 7* capsule. (*NASA*)

called *Friendship 7* for five hours, orbiting our planet three times. He later became a U.S. Senator, representing Ohio.

## Who was the **oldest person** to **fly into outer space**?

In 1998 John Glenn became the oldest person ever to go into space when, at the age of 77, he flew in the space shuttle *Discovery*.

## Who was the **first** person to **fly in space twice**?

Virgil "Gus" Grissom (1926–1967) first flew into space in July 1961, on the second suborbital Mercury mission. It was officially called *Mercury-Redstone 4,* but the spacecraft was more popularly known as *Liberty Bell 7*. In March 1965, he piloted the first manned flight for the Gemini program, *Gemini 3* (nicknamed the *Molly Brown*), and orbited Earth three times with co-pilot John Young (1930–). Grissom thus became the first man to fly into space twice.

## Who was the **first person** to **walk in space**?

Soviet cosmonaut Alexei Leonov (1934–) was the first person to travel in outer space outside of a spacecraft. On March 18, 1965, he floated for twelve minutes outside of his vessel, *Voskhod 2*. Leonov's historic mission was the tenth piloted space mission in history, and the sixth for the Soviet Union.

On *Voskhod*'s second orbit around Earth, Leonov put on a spacesuit and a backpack containing an oxygen tank, and entered the spacecraft's airlock. When the entrance to the vessel was resealed, Leonov opened the outer hatch and climbed out. He floated 17 feet (5.3 meters) away from the spacecraft, the full length of his safety line. He landed on top of the craft, where he remained for a few minutes before pulling himself back to the hatch. Leonov then found out that his spacesuit had ballooned out in several places, making it impossible for him to fit back inside the hatch. Fortunately, he quickly solved the problem by releasing some air from the pressurized suit.

## Who was the **first American astronaut** to **walk in space**?

A few months after Leonov became the first person to walk in space, astronaut Ed White (1930–1967) undertook the first American spacewalk on June 3, 1965. For 21 minutes, White remained outside the *Gemini 4* capsule attached to a tether, while his shipmate James McDivitt (1929–) looked on from inside the capsule.

## Who were the **first African American** astronauts in **outer space**?

Dr. Guion "Guy" Bluford, Jr. (1942–) achieved the rank of colonel in the U.S. Air Force as a fighter pilot and earned his Ph.D. in aerospace engineering in 1978. He joined NASA's astronaut corps soon after, and on August 30, 1983, he became the first African American man in space when he served as a mission specialist on the space shuttle *Challenger*. Bluford then made three more shuttle flights, in October 1985, April 1991, and December 1992.

Dr. Mae Carol Jemison (1956–) earned her M.D. in 1981 from Cornell Medical School in New York. After studying and working in Cuba, Kenya, and a Cambodian refugee camp in Thailand, and serving as a Peace Corps volunteer practicing medicine for two years in Sierra Leone and Liberia, she was accepted into the astronaut corps in 1987. On September 12, 1992, as a member of the crew of the space shuttle *Endeavour,* Jemison became the first African American woman in outer space.

## Who were the first **Asian American** astronauts in **outer space**?

Ellison Shoji Onizuka (1946–1986) was born in Hawaii, earned a master's degree in aerospace engineering, and served in the U.S. Air Force as a flight test engineer and test pilot, reaching the rank of lieutenant colonel. In 1978 NASA selected him for the astronaut corps, and he worked on a number of space shuttle missions on the ground. On January 24, 1985, as a member of the crew of the space shuttle *Discovery,* Onizuka became the first Asian American man in outer space. Tragically, on his second mission into space, he and his fellow shuttle crew members were killed aboard the space shuttle *Challenger* on January 28, 1986.

Dr. Kalpana Chawla (1961–2003) was born in Karnal, Haryana, India, and earned a Ph.D. in aerospace engineering in 1988 at the University of Colorado. She held a certified flight instructor rating and a commercial pilot's license for numerous kinds of aircraft. She became a naturalized U.S. citizen in 1990, and joined the NASA astronaut corps in 1995. On November 19, 1997, as a member of the crew of the space shuttle *Columbia,* Chawla became the first Asian American woman in outer space. Sadly, on her second mission into space, she and her fellow shuttle crew members were killed aboard the space shuttle *Columbia* on February 1, 2003.

# EARLY SOVIET PROGRAMS

## What was the *Vostok* like?

*Vostok,* Russian for "east," was a small, relatively simple spacecraft, consisting of a cabin and an instrument module. The spherical, 7.5-foot-diameter cabin was large enough to accommodate only one person. The outside of the cabin was coated with a protective heat shield. Communication antennae extended from the top of the cabin, and nitrogen and oxygen tanks for life support were stored beneath it. The instrument module, containing a small rocket and thrusters, was strapped to the cabin with steel bands.

## What was the **Voskhod program**?

Voskhod, Russian for "sunrise," was the Soviet Union's second series of piloted spacecraft. It was similar in design to its predecessor, the Vostok series, except that it could hold three humans at a time, rather than just one. Voskhod was created as a stopgap craft in the space program to keep the Soviet manned space program moving forward as delays in the Soyuz program mounted. As such, the spacecraft was a bit rough around the edges; the cosmonauts sat on small couches, there were no ejection seats or emergency escapes, and there was so little room in the cabin that the three cosmonauts could not even wear spacesuits. Fortunately, though the Voskhod was fraught with risk, no mishaps occurred for the lifetime of the progam.

*Voskhod 1* cosmonauts (from left to right) Vladimir Komarov, Boris Yegorov, and Konstanin Feoktistov. (*Peter Gorin*)

## What were the first **Voskhod missions**?

Following one unmanned test flight, *Voskhod 1* was launched on October 12, 1964, with three men aboard. It successfully returned to Earth after one day. *Voskhod 2* was launched on March 18, 1965, and cosmonaut Alexei Leonov (1934–) made the first-ever human spacewalk during that flight. As

> ## What was the *Soyuz 11* tragedy, and how did it affect the Soviet space program?
>
> There have been dozens of manned launches of the Soyuz series vehicles, almost all of them successful. *Soyuz 11,* however, ended tragically. It launched on June 6, 1971, and completed its mission to rendezvous with the *Salyut 1* space station. During the crew's descent to Earth, however, a valve opened unexpectedly, allowing all the air in the cabin to escape. All three cosmonauts onboard suffocated. After this mishap, a number of changes were made to the Soyuz crafts, and the number of cosmonauts on any mission was reduced to two, allowing each occupant to wear a pressurized space suit during launch, docking, and re-entry.

Leonov and fellow cosmonaut Pavel Belyayev were preparing to return to Earth, however, they noticed that their ship was pointed in the wrong direction. It took them another orbit to turn the spacecraft around, causing them to alter their landing site. The two crewmates parachuted to Earth in a remote region of the Ural Mountains, and they spent two days in the forest before rescue teams reached them. No further Voskhod missions ever took place, probably because the Soyuz program was close enough to completion that the Soviets decided to focus their energies there.

## What was the **Soyuz program**?

The Soyuz (Russian for "union") program is the longest-running Soviet (and later Russian) space mission program to date. The program was originally intended for missions to the Moon. The head of the Soviet space program, Sergei Korolëv (1906–1966), designed a series of three Soyuz spacecraft for this purpose in the early 1960s. In 1964, however, the Soviets decided to use a more powerful Proton rocket for Moon flights. They scaled back the Soyuz program to a series of spacecraft to be used for Earth-orbiting missions.

## What were the first **Soyuz missions**?

*Soyuz 1,* launched on April 23, 1967, was comprised of three sections: an orbital module, a descent module, and a compartment containing instruments, engines, and fuel. Unfortunately, the mission was plagued with problems and ended in tragedy as its parachutes failed to open just before landing. The spacecraft crashed to Earth, and cosmonaut Vladimir Komarov (1927–1967) was killed. Future Soyuz missions were more successful. *Soyuz 3* carried cosmonaut Georgi Beregovoy (1921–1995) into space and back safely. *Soyuz 4* and *Soyuz 5* successfully launched in January 1969, and the cosmonauts, Aleksey Yeliseyev (1934–) and Yevgeni Khrunov (1933–2000), each performed space-walks and switched vehicles, accomplishing the first-ever crew transfer in space.

The *Soyuz* cosmonauts (standing, left to right) Viktor Gorbatko, Anatoliy Filipchenko, and Vladislav Volkov, (seated, left to right) Valeriy Kubasov, Georgiy Shonin, Vladimir Shatalov, and Aleksey Yeliseyev. (*Peter Gorin*)

## What are the **Soyuz launch vehicles**?

*Soyuz* is also the name given to a series of launch vehicles, advanced versions of which are still in use today. The European Space Agency's *Venus Express* spacecraft, for example, was launched by a Soyuz rocket in 2005.

## What was the **Luna program**?

The Luna program was run by the Soviet Union between 1959 and 1976, with the purpose of exploring the Moon and its surroundings using space probes. A series of 24 Luna probes achieved a number of milestones in unpiloted space exploration, including orbiting, photographing, and landing on the Moon.

## What are some **accomplishments** of the **Luna program**?

In 1959 *Luna 1* was the first spacecraft to fly by the Moon. *Luna 2* was launched on September 12, 1959, and crash-landed onto the lunar surface, becoming the first human-made object to reach the Moon. A few months later, *Luna 3* took the first pictures of the far side of the Moon. In February 1966, *Luna 9* was the first human-made object to make a soft landing on the Moon. The ball-shaped space probe contained a television camera, which transmitted footage of the moonscape around it. In September 1970, *Luna 16* became the first of four probes to collect lunar soil samples robotically and return them to Earth. Between November 1971 and January 1973, Luna probes placed two remote-controlled, lunar roving cars on the Moon. *Lunakhod 1* and *Lunakhod 2* cruised over the lunar terrain, taking photographs and measuring the chemical composition of the soil.

# EARLY AMERICAN PROGRAMS

## What was the **Mercury program**?

The Mercury program ushered in the era of American space flight. It was begun in 1959 by the newly formed National Aeronautics and Space Administration (NASA).

## What were **Mercury capsules** like?

The Mercury spacecraft had a bell-shaped capsule a little less than nine feet tall (2.74 meters) and six feet (1.8 meters) wide. It was so small that it could accommo-

## Who was Ham?

The Mercury program included a series of unpiloted test flights that was followed by a test flight, in January 1961, carrying a chimpanzee named Ham. When Ham returned safely, Mercury was deemed ready for a human pilot.

date only a single astronaut at a time. The astronaut entered through a square hatch in the side of the capsule and sat on a chair that had been specially shaped to fit his body. Directly in front of the chair was the control panel. The base of the capsule was enclosed in a heat shield designed to withstand the searing heat of re-entry into Earth's atmosphere. Just before landing, the shield gave way to an inflated cushion, and parachutes sprang from the top of the capsule.

## Who were the **"Mercury Seven"**?

The "Mercury Seven" were the first people to be selected to the U.S. astronaut corps. They were Walter M. Schirra Jr. (1923–2007), Donald K. "Deke" Slayton (1924–1993), John H. Glenn Jr. (1921–), M. Scott Carpenter (1925–), Alan Bartlett Shepard Jr. (1923–1998), Virgil "Gus" Grissom (1926–1967), and L. Gordon Cooper Jr. (1927–). All became national heroes.

## What were the **accomplishments** of the **early Mercury program**?

Six piloted missions were launched from 1961–1963. The early Mercury vessels were launched into space by Redstone rockets; later Mercury craft were launched by Atlas rockets. These short Mercury flights led to the longer, more complex Gemini flights of the mid-1960s, and finally to the Apollo program which had its last flight in 1972.

## What was the **Gemini program**?

The Gemini program was the second phase in American spaceflight. In all, 12 Gemini spacecraft were launched between April 1964 and November 1966. On those missions, astronauts learned space skills like docking with other vessels and conducting spacewalks, setting new records for endurance and altitude.

Astronaut John Glenn prepares himself for the Mercury-Atlas 6 mission. (*NASA*)

211

Those flights solved a number of spaceflight problems and paved the way for the Apollo program.

## What were **Gemini spacecraft** like?

The Gemini spacecraft were larger than the Mercury spacecraft and could hold two astronauts. The Gemini vessels also had maneuvering thrusters and were capable of changing orbits, linking with other spacecraft, and precisely controlling their re-entry and landing. Gemini was a very successful program, though *Gemini 8* came close to disaster.

## What were some **accomplishments** of the **Gemini program**?

Among the many achievements of the Gemini mission were the first American spacewalks; altitude records of more than 850 miles (1,370 kilometers) above sea level; a then-record 14-day mission in space; and the first docking of two spacecraft.

# THE APOLLO MISSIONS

## How many **lunar missions** have there been?

Since 1958, more than 60 space vehicles have been launched toward the Moon. Most of them have been unpiloted. This includes spacecraft that have flown past the Moon; those that have gone into orbit around the Moon, sending information back to Earth for months or years; and those that have missed their target altogether and ended up orbiting the Sun. Some Moon-bound spacecraft have crash-landed on the lunar surface or descended to a soft landing, collecting soil samples and other scientific data. The most celebrated of all lunar vehicles have been the piloted Apollo missions.

## What were the **lunar exploration programs** launched by the United States before the **Apollo period**?

The U.S. lunar exploration of the Moon consisted of a number of programs, culminating in the manned Apollo program that brought people to the Moon for the first

## How many flights were made in the Apollo program?

The first Apollo program launch occurred two months after the final Gemini mission. It was the culmination of the U.S. space program's decade-long quest to send a human being to the surface of the Moon and return safely from the journey. The first space mission to land humans on the Moon was *Apollo 11,* which was launched on July 16, 1969 and landed Neil Armstrong (1930–) and Buzz Aldrin (1930–) on the surface on July 20. Six more Apollo missions were sent to the Moon after that—one unsuccessful, the others successful. After *Apollo 17* returned in 1972, three more trips to the Moon were cancelled due to budgetary constraints and a change in national space exploration priorities.

time in human history. The first few probes in the Pioneer series culminated in the lunar fly-by of *Pioneer 4* in March 1959. The Ranger program sent a total of nine space probes to the Moon; the last three members of the fleet—*Rangers 7, 8,* and *9*—were launched in 1964 and 1965; they transmitted detailed pictures of the lunar surface before crash landing.

Between 1965 and 1968, the United States deployed a dozen more space probes to the Moon. The Lunar Orbiter vessels went into orbit around the Moon, while the Surveyor spacecraft soft-landed on the lunar surface. These spacecraft collected important information that would assist in planning the route and landing sites of the human-piloted Apollo program.

## What did the **Apollo missions** to the Moon **accomplish**?

The Apollo program was the focus of the United States space program from 1967 to 1972. Beginning with *Apollo 11,* which landed on the Moon on July 20, 1969, Apollo spacecraft landed 12 men on the Moon's surface. Aside from gathering tremendous amounts of new information about the Moon, and bringing back 842 pounds (382 kilograms) of Moon rock, the Apollo program showed conclusively that it was possible for humans to set foot on an object in the universe other than Earth, proving that outer space is not a barrier, but a frontier.

## What were the **Apollo spacecraft** like?

The Apollo spacecraft consisted of three parts: a command module, where the astronauts would travel; a service module, which contained supplies and equipment; and a lunar module, which would detach to land on the Moon. In all, a total of 15 Apollo spacecraft were produced—three designed for unpiloted missions and 15 for piloted missions. The Apollo missions were launched using the Saturn V rocket, designed by Wernher von Braun and still the most powerful rockets ever successfully operated.

The crew of the *Apollo 11* were, from left to right, Commander Neil A. Armstrong, Command Module Pilot Michael Collins, and Lunar Module Pilot Edwin E. Aldrin, Jr. (*NASA*)

### What did the **early Apollo missions** achieve?

Three unpiloted Apollo missions were flown in 1967 and 1968 as test flights. The first successful piloted mission, *Apollo 7,* was launched in October 11, 1968; three astronauts orbited Earth for eleven days. Two months later, the crew of *Apollo 8* became the first humans to escape Earth's gravitational field and orbit the Moon. *Apollo 9* and *Apollo 10* had flights in early 1969, and were used for final preparation runs for the landing mission in July.

### Where did the **famous photo** of **planet Earth** come from during the Apollo missions?

One of the most famous photographs in human history—the image of Earth rising over the lunar horizon—was taken by the *Apollo 8* crew as it orbited the Moon.

### Who were the **astronauts** onboard *Apollo 11*?

American astronauts Neil Armstrong (1930–), Edwin Eugene "Buzz" Aldrin (1930–), and Michael Collins (1930–) were aboard *Apollo 11* when it made its historic trip to the Moon. Armstrong was born in Ohio and became a fighter pilot for the U.S. Navy. He earned a master's degree in aerospace engineering, and joined the astronaut corps in 1962. Aldrin grew up in Montclair, New Jersey, served as a fighter pilot for the U.S. Air Force, and earned his doctorate in astronautics before join-

**Exactly what did Neil Armstrong and Buzz Aldrin say when they first set foot on the Moon?**

When Neil Armstrong first planted his foot on the Moon's surface, he said, "That's one small step for [a] man, one giant leap for mankind." Armstrong said afterward that he had definitely meant to say "a man," instead of just "man," but either the word caught in his throat, or the audio transmission from the Moon dropped out at just that moment. In any case, Armstrong has stated that he prefers to have the "a" included in parentheses to acknowledge the possibility that either one may be correct. Buzz Aldrin's first words on the Moon were unequivocal: "Magnificent desolation," he said.

ing the NASA astronaut corps in 1963. Collins attended the U.S. Military Academy, then served as a pilot in the Air Force, eventually earning the rank of major general. Like Aldrin, he joined the NASA astronaut corps in 1963. All three of them had one spaceflight experience before going to the Moon—Armstrong on *Gemini 8,* Collins on *Gemini 10,* and Aldrin on *Gemini 12.*

## What happened when *Apollo 11* reached the Moon?

When *Apollo 11* reached the Moon, Neil Armstrong and Buzz Aldrin headed for the surface in the lunar module, nicknamed "Eagle," while Michael Collins stayed in the command module in orbit around the Moon. After moving across the Sea of Tranquility in the lunar module, looking for a safe place to set down, Armstrong finally landed the Eagle with less than a minute of fuel to spare. Armstrong and Aldrin then planted the American flag on the Moon's surface, took photographs, held a telephone conversation with President Richard Nixon, set up several science experiments, and collected rocks and soil samples. Before they left the Moon's surface three hours later, they left behind a plaque that read, "Here men from the Planet Earth first set foot upon the Moon. July 1969 A.D. We came in peace for all mankind."

## What happened on the **other missions** to the **Moon** after *Apollo 11*?

Six more missions to the Moon occurred after the historic flight of *Apollo 11.* One of them almost ended in tragedy: *Apollo 13* was more than halfway to the Moon when an explosion ruptured the oxygen tanks and destroyed most of the ship's systems. Through courage, luck, and incredible hard work on Earth and in space, the spacecraft managed to loop around the Moon, using the flyby as a gravitational slingshot to fling the ship and its crew back to Earth—shaken, but alive.

The other five missions went off without a hitch. Fifteen more astronauts went to the Moon in those missions, and 10 of them set foot on the surface. They explored, conducted geological and astronomical experiments, collected lunar rocks and soil, rode in a "Moon buggy," and more.

**What sport did Alan Shepard play
on the Moon during the *Apollo 14* mission?**

After *Apollo 14* landed on the Moon, astronaut Alan Shepard hit a golf ball—which, according to him, went "miles and miles and miles" thanks to the Moon's weaker gravitational pull.

## What happened after the **last Apollo mission**?

After the last mission to the Moon on *Apollo 17,* the United States government cancelled the remaining three scheduled Moon visits because of budgetary issues. Since then, there have been no manned landings on the Moon.

## Who are **all the people** who have **walked on the Moon** and **when** did they do it?

The following table lists all the astronauts who have set foot on the Moon.

### Humans Who Have Walked on the Moon

| Name | Moon Mission | Date of Moon Landing |
| --- | --- | --- |
| Neil Armstrong | *Apollo 11* | July 20, 1969 |
| Buzz Aldrin | *Apollo 11* | July 20, 1969 |
| Pete Conrad | *Apollo 12* | November 19, 1969 |
| Alan Bean | *Apollo 12* | November 19, 1969 |
| Alan Shepard | *Apollo 14* | February 5, 1971 |
| Edgar Mitchell | *Apollo 14* | February 5, 1971 |
| David Scott | *Apollo 15* | July 31, 1971 |
| James Irwin | *Apollo 15* | July 31, 1971 |
| John W. Young | *Apollo 16* | April 21, 1972 |
| Charles Duke | *Apollo 16* | April 21, 1972 |
| Eugene Cernan | *Apollo 17* | December 11, 1972 |
| Harrison Schmitt | *Apollo 17* | December 11, 1972 |

## What was the **Apollo-Soyuz** mission and why was it famous?

After the final Apollo flight to the Moon, the next spacecraft, *Apollo 18,* was used for another historic event. On July 15, 1975, the Soviet Union's *Soyuz 19* spacecraft was launched with cosmonauts Alexei Leonov (1934–) and Valeriy Kubasov (1935–) onboard. Seven hours later, *Apollo 18* took off, carrying astronauts Thomas Stafford (1930–), Vance Brand (1931–), and Donald "Deke" Slayton (1924–1993), as well as a special docking module designed to fit *Soyuz* at one end and *Apollo* at the other, with an airlock chamber in between. That evening, the two spacecraft successfully rendezvoused with one another and docked together. The Americans entered *Soyuz* and the two crews shook hands on live television.

The two spacecraft remained docked for two days, during which time they carried out joint astronomical experiments. After separating, *Soyuz 19* returned to

## Why do some people believe that humans have not actually visited the Moon?

**A**ccording to psychologists, the idea of having knowledge of secrets, conspiracies, or other "special" information is alluring to people. That is why, every few years or so, a television show or Internet rumor recirculates about the Moon landings being a hoax. Some people are willing to believe them, even though those shows and rumors are false.

The thought of faked Moon landings does indeed fire our imaginations. But the reality—that thousands of people, working together for years, spending billions of dollars, conducting unprecedented research, and creating remarkable feats of science and engineering, actually landed people on the Moon and brought them safely back to Earth—is far more fascinating and awe-inspiring. The work of the Apollo project—and of the space program overall—is thoroughly documented, with thousands of hours of recordings and millions of pages of paper, and are widely available for everyone to see and study.

Earth, while *Apollo 18* stayed in orbit for three more days. Both vessels landed safely. Many people consider the Apollo-Soyuz Test Project—the first joint venture in space between the United States and the Soviet Union—the beginning of the end of the hostile "space race" that started in the 1950s, and the start of the modern era of international cooperative human activity in space.

## What **spacecraft** have **explored** the Moon **after the Apollo program** ended, and what have they found?

Exploration of the Moon has been very slow since the early 1970s, when the Apollo program ended. In 1990 the Japanese twin Muses-A space probes reached orbit around the Moon, but failed to transmit any data. The *Clementine* probe was launched by the United States in 1994, and surprisingly detected signs of water ice mixed in with rock near the Moon's south pole. A follow-up mission, called the Lunar Prospector, was launched in January 1998. By March 1998, data from the spacecraft suggested that this subterranean ice might be present in large quantities at both lunar poles. However, when the Lunar Prospector mission was ended in July 1999, the spacecraft was controlled-crash-landed at the lunar south pole. No ice was detected on the surface. The scientific debate continues, and it is important because the presence of water ice on the Moon could significantly influence future human colonization of the Moon.

## Since there is **no air or wind on the Moon**, how come the **flags** planted by the Apollo astronauts **stand straight out**?

The American flags planted on the Moon are attached on their side to a vertical flagpole, and they are also attached at the top by a horizontal crossbar. They were hung in such a way that the crossbars are not generally visible in the photographs taken by the Apollo astronauts, although the crossbar can be seen if one looks carefully. Also, when a flag is planted on the Moon, the vibrations traveling through the flag-pole will temporarily cause the cloth flag to shake and wave. Those motions will last for quite a while because there is no air resistance on the Moon to slow them down.

# EARLY SPACE STATIONS

### What was the **Salyut program**?

On April 19, 1971, the Soviet Union launched *Salyut 1,* the world's first space station. It was designed to accommodate three cosmonauts for three to four weeks. In all, the Soviet space program operated seven *Salyut* space stations between 1971 and 1991. These space stations helped scientists and spacecraft designers understand some of the challenges and possibilities of extended stays in space.

### What were the **Salyut space stations** like?

*Salyut 1* was built in the shape of a tube, 47 feet (14 meters) long and 13 feet (four meters) across at its widest point, and it weighed 25 tons. Four solar panels extended from its body like propellers, providing the station's power. It contained a work compartment and control center, a propulsion system, sanitation facilities, and a room for scientific experiments. It was used only once by a three-man crew for 24 days. That crew died tragically during their descent to Earth in the *Soyuz 11* spacecraft on June 30, 1971. The future Salyut space stations were built approximately in the same way, with improvements and modifications. *Salyut 4* had a different distribution of solar panels and a solar telescope at one end, and *Salyut 6* and *Salyut 7* had two docking ports instead of one. *Salyut 7* was also a prototype "modular" space station, with numerous pieces that could be added on after launch to increase the station's size and capabilities.

### What **milestones** were achieved by the **Salyut stations**?

*Salyut 6* was launched on September 29, 1977, and remained in orbit until July 1982. During that time, it received numerous sets of cosmonauts, as well as supplies carried by unpiloted *Progress* spacecraft. The longest stay on *Salyut 6* by any crew was 185 days. *Salyut 7* was launched on April 19, 1982, and also hosted many delegations of cosmonauts. The longest visit lasted 237 days, and the last time the space station was occupied was in March 1986, when cosmonauts living aboard the space station *Mir* visited the spacecraft. They spent six weeks there before returning to their larger, more permanent home. *Salyut 7* burned up in Earth's atmosphere on February 7, 1991.

### What was *Skylab*?

*Skylab* was a space station operated by the United States from 1973 to 1979. The two-story *Skylab* was much larger than its contemporary *Salyut* space stations. It

*Skylab* was the first U.S. space station and had a highly successful run from 1973 through 1979. This 1974 photo shows the gold shield erected to protect the station's workshop after the original micrometeroid shield was lost. (*NASA*)

was 118 feet (36 meters) long, 21 feet (6.4 meters) in diameter, and weighed 80 tons. It contained a workshop, living quarters for three people, a module with multiple docks, and a solar observatory. At an altitude of 270 miles (440 kilometers), *Skylab* holds the record for the largest orbital distance from Earth's surface of any human-occupied space station.

## What happened when *Skylab* was launched?

Almost immediately after its launch on May 14, 1973, *Skylab* encountered problems. The space station's meteoroid shield, thermal shield, and one of its solar panels were lost, while the second solar panel was jammed. The station's power system was also damaged. Eleven days after *Skylab*'s launch, its first crew arrived, repairing most of the damage and restoring power to the station. The crew remained for 28 days and carried out a number of scientific experiments before returning to Earth.

## How did the Skylab program **end**?

In all, three crews lived on *Skylab* from 1973 to 1974. They stayed aboard for 28, 59, and 84 days respectively, and conducted a great deal of scientific research, especially solar studies and biomedical studies of the effects of weightlessness on animal and plant life.

After the third crew left, the station was placed in a parking orbit and expected to last there for at least eight years. Unfortunately, unexpectedly high atmospheric

219

drag pulled the spacecraft into a lower orbit much more quickly than originally calculated. A plan was made for a space shuttle to dock with *Skylab* in 1979 and ferry the station to a higher orbit; but the shuttle program experienced years of delays and was not ready to launch until 1981. Another plan was made to send an unmanned spacecraft to save *Skylab,* but it was not funded by the U.S. government. On July 11, 1979, *Skylab* fell back to Earth, scattering fragments from the middle of the Indian Ocean all the way to Australia.

## What was *Mir*?

*Mir,* meaning "peace" in Russian, was an orbiting space station operated by the Soviet Union (and its successor government, Russia) from 1986 to 2001. *Mir* was launched in self-contained, attachable pieces called modules, and was literally put together in space one module at a time. The first (core) module was launched on February 19, 1986, from Baikonur Cosmodrome in Kazakhstan. By 1996, when the seventh and last module was installed, *Mir* had become a multi-spoked cylinder more than 100 feet long, had a mass of more than 120 tons, and had more than 10,000 cubic feet of living space.

## How was *Mir* configured?

The main body of the *Mir* space station consisted of four areas: a docking compartment, living quarters, a work area, and a propulsion chamber. The docking compartment contained television equipment, the electric power supply system, and five of the vessel's six docking ports. The work area was the spacecraft's nerve center and contained the main navigational, communications, and power controls. At one end of the station, the unpressurized propulsion compartment contained the station's rocket motors, fuel supply, heating system, and the sixth docking port to receive unpiloted refueling missions.

As modules were added to the station, *Mir* continued to gain mass and functionality. An observatory module with ultraviolet, X-ray, and gamma-ray telescopes was added in 1987; a module with two solar panel arrays and an airlock was added next in 1989; and a scientific module was added in 1990. In 1995 two more modules were added, one of which was a docking module carried to the station by the space shuttle *Atlantis;* and a remote Earth sensing module was added in 1996.

## How did the *Mir* mission **end**?

By 1997 the *Mir* space station had more than doubled its original warrantied lifetime of five years. The years of service began to take its toll on the vessel's systems, and things began to break down. By June 1997, crises were becoming almost commonplace: a fire, a cooling system that leaked antifreeze, a faulty oxygen processing system, a collision with a space cargo ship, a computer crash, and more plagued the station. On August 28, 1999, the station's crew was returned to Earth—the first time in nearly 10 full years that *Mir* was left unoccupied.

On April 4, 2000, a crew of two cosmonauts returned to *Mir* to assess the ship's condition and future prospects. After they left on June 16, no further visits to the station were made. To ensure the safety of people living on Earth, an unmanned rocket was sent to the station. Flight controllers then used that rocket to bring *Mir* down into the atmosphere and de-orbit the vessel. On March 23, 2001, *Mir* burned up upon re-entering Earth's atmosphere, lighting up the skies over the Fiji Islands and scattering debris harmlessly across the southern Pacific Ocean.

# THE SPACE SHUTTLE

## What is the **Space Shuttle program**?

The Space Transportation System (STS), better known as the Space Shuttle program, is NASA's primary piloted space program. The space shuttle was designed in the 1970s as a half-spacecraft, half-airplane, reusable system that could frequently ferry people and cargo into low Earth orbit and back. There have been more than 120 space shuttle flights. Although it has had a checkered history marked with cost overruns and two terrible tragedies, the shuttle program has also created tremendous successes in human spaceflight and helped scientists and engineers understand what life in space—and the process of coming and going from space to Earth and back—may someday be like.

## How is the **space shuttle configured** and **operated**?

The space shuttle system consists of a main liquid fuel tank, two solid rocket boosters (SRBs), and the shuttle orbiter. When the shuttle is launched, the orbiter and SRBs are attached to the main fuel tank, and the tank fuels the orbiter's three main engines. A    221

The space shuttle *Atlantis* leaves dock from the *Mir* space station. (*NASA*)

few minutes after launch, the SRBs exhaust their fuel, detach from the main tank, and fall into the ocean; a parachute system slows their fall, and they are recovered for use in future launches. The main tank and orbiter stay together until low Earth orbit is achieved. When the main tank is empty, it is detached as well. It cannot be recovered and generally burns up in the atmosphere. The orbiter, with astronauts aboard, then goes on to complete the mission. This 184-foot (56-meter) long vessel contains engines, rocket boosters, living and work quarters for up to eight crew members, and a cargo bay large enough to hold a large school bus. It also has wings and is aerodynamically designed to be able to glide back to Earth from orbit, landing like an airplane on any runway long enough to accommodate a commercial jumbo jet.

### How **many space shuttles** are there?

Six shuttles were built. The first shuttle orbiter, *Enterprise,* was constructed for test purposes and was never launched into orbit. It proved capable, however, of lifting

## Which space shuttles have been tragically lost?

The *Challenger* was destroyed during launch on January 28, 1986, with all seven crew members lost. The *Columbia* disintegrated after reentering the atmosphere on February 1, 2003; again, all seven crew members perished. The three remaining active shuttle orbiters are the last of their kind. The shuttle program is scheduled to be retired within a few years, and no more will be built.

off and gliding down to a safe landing. The first shuttle orbiter to be launched into space was *Columbia,* piloted by astronauts John Young (1930–) and Robert Crippen (1937–), which was launched for the first time on April 12, 1981, and landed safely on April 14. It was followed by *Challenger* on April 4, 1983; *Discovery* on August 30, 1984; *Atlantis* on October 3, 1985; and *Endeavour* on May 7, 1992.

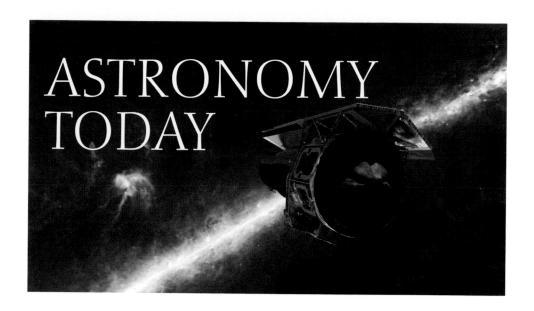

# MEASURING UNITS

### What is an **astronomical unit**?

An astronomical unit, or AU, is defined as the average distance between Earth and the Sun. It is just about equal to 93,000,000 miles (149,600,000 kilometers). Most astronomers approximate 1 AU = 150,000,000 kilometers. For comparison, Mercury is about 0.4 AU from the Sun; Pluto is about 50 AU from the Sun; and the Alpha Centauri system, which contains the stars closest to the Sun, is about 270,000 AU away.

### How do **astronomers measure sizes and distances** in the universe?

Since tape measures are pretty inconvenient measuring tools in astronomy, other methods are used. Geometric methods like parallax and standard candles, such as Cepheid variables, are the most common ways to make such measurements. Along with the standard terrestrial units of length, like meters or miles, specialized units have been developed to make cosmic measurements. These include astronomical units (AU), light-years, and parsecs.

### How was the **astronomical unit** first measured?

Italian astronomer Gian Domenico Cassini (1625–1712), who is famous for studying the rings of Saturn, was the first astronomer to make a nearly accurate measurement of the length of the astronomical unit. Cassini first measured the parallax of Mars, based on his own observations made in Paris and those of his colleague Jean Richer in South America. With this information he was able to calculate the distance from Earth to Mars, and from that the distance from Earth to the Sun. Cassini's measurement was slightly low at about 87 million miles (140 million kilometers), but he was off by less than ten percent of the correct value: 93 million miles (149.6 million kilometers).

## What is a **light-year**?

A light-year is the distance a beam of light travels through the vacuum of space in one Earth year. Since light in a vacuum travels at about 186,000 miles (300,000 kilometers) per second, and there are about 31,500,000 seconds in a year, a light-year is about 5.88 trillion miles (9.47 trillion kilometers).

## What is a **parsec**?

The word "parsec" is constructed from "parallax arcsecond." Taking the width of Earth's orbit around the Sun as a baseline, an object one parsec away would provide a parallax measurement of one arcsecond. That distance is about 19 trillion miles (31 trillion kilometers), or about 3.26 light-years.

## What is a **kiloparsec** and a **Megaparsec**?

A kiloparsec (or kpc for short) is 1,000 parsecs, and a Megaparsec (or Mpc for short) is 1,000,000 parsecs. For reference, the typical separation between stars in the disk of the Milky Way galaxy is about a few parsecs; the diameter of the disk of the Milky Way is about 30 kiloparsecs; and the distance between the Milky Way and the Andromeda galaxy is about 0.7 Megaparsec.

## What is **astrometry**?

Astrometry is the astronomical measurement of positions (positional astronomy) and motions (dynamical astronomy). It is very important to know how objects in the universe are moving—or, conversely, not moving. Astrometry of near-Earth asteroids, for example, can help us determine if any objects in the solar system are likely to strike our planet. Astrometry of stars helps us understand how our solar system moves within the Milky Way galaxy. Furthermore, astrometry is very important in establishing reliable frames of reference, in time as well as space, for both scientific and everyday use. The United States Naval Observatory, for example, continually measures and records the movements of the Sun, Moon, planets, and stars; the data are passed on to the Nautical Almanac Office, which together with the British government publishes the *Astronomical Almanac*. The annual almanac is used as a daily reference for navigation, surveying, and science.

## What is **parallax** and how does it work?

The general idea of parallax is to use triangulation to measure distances. When looking at an object from two different vantage points, the object appears to shift its position relative to the background. For astronomical applications, the position of Earth shifts by up to 186 million miles (300 million kilometers) as Earth orbits the Sun. So it is possible to view distant objects, such as stars, at two different vantage points. The measure of the amount of apparent change in position of that

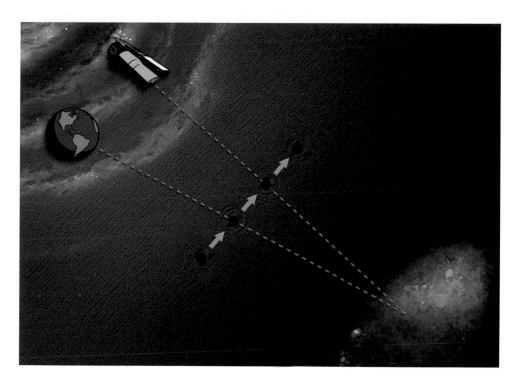

By viewing objects in space from different positions, such as from the Earth and from a space telescope, astronomers can determine their distance. (*NASA/JPL-Caltech/T. Pyle (SSC)*)

object is its parallax. Once the parallax is known, it is possible to calculate the distance to that object.

## What is a **standard candle**?

A standard candle is an object that has the same luminosity, or energy output, wherever it appears in the universe. Imagine if every blinking red-colored flashlight bulb were exactly 100 watts, no matter where it is seen; in that case, viewing this light at night can tell a person how far away that flashlight is by measuring how bright the red light appears.

Unfortunately, not many bright objects in the universe are standard candles. Red stars, for example, can have very different luminosities. It is extremely important, therefore, to find very luminous objects that serve as standard candles, so we can measure distances to faraway objects in the universe that are too distant to be measured using parallax.

## Who **discovered** the kind of standard candle called **Cepheid variables**?

American astronomer Henrietta Swan Leavitt (1868–1921) worked at the Harvard College Observatory in Cambridge, Massachusetts. In 1904, Leavitt noticed that a particular star in the constellation Cepheus would regularly change its brightness. Careful study showed that the star varied its brightness in a predictable, "saw-tooth" pattern. Eventually, other variable stars with this same saw-tooth pattern were found and were named Cepheid variables after the first star of this type that was discovered.

In 1913 Leavitt and the Danish astronomer Ejnar Hertzsprung (1873–1967) worked together to deduce that Cepheid variables varied in a very specific way: the time it takes a Cepheid variable to go through one cycle of brightness variation—its period—is mathematically related to the peak luminosity of the star. This kind of "period-luminosity relation" meant that it was possible to use Cepheid variables as standard candles: to know the luminosity of a Cepheid, just measure its period of variability, and from that, the distance to the star or the object it is in.

### How did **Edwin Hubble** use **Cepheid variables** to **measure the universe**?

In the early twentieth century, it was not yet known whether so-called "spiral nebulae" were inside our Milky Way galaxy or outside of it. In 1924, American astronomer Edwin Powell Hubble (1889–1953) began a study of spiral nebulae, using the 100-inch Hooker Telescope at Mount Wilson Observatory in California. Over many months, Hubble identified hundreds of Cepheid variable stars in the largest spiral nebula toward the constellation Andromeda. Using the period-luminosity relation of Cepheid variables, he showed that the Andromeda spiral nebula is at least about one million light-years away—a far larger distance than the size of the Milky Way. Also, at that distance Andromeda would have to be many thousands of light-years across to be visible. Thus, Hubble proved that the Andromeda spiral nebula is in fact the Andromeda galaxy, and that the universe contains not just one, but many galaxies that are millions of light-years apart.

# TELESCOPE BASICS

### What is a **telescope**?

Generally speaking, a telescope is an instrument that gathers light from distant sources in such a way that an image can be produced. The first telescopes were made with glass lenses attached to handheld cylinders or tubes. Today, telescopes

Telescopes have advanced considerably since the days of Galileo. A modern child's telescope can see objects like Saturn and the Andromeda galaxy as well as, or even better than early telescopes. (*iStock*)

are made in many different ways, and used together with all manner of scientific instruments, to study the universe near and far.

## Who **invented** the **telescope**?

It is thought that, in the early 1600s, a Dutch optician named Hans Lippershey (c. 1570–c. 1619) built the first telescope. Many people, however, were converging on this new technology around that time. By 1609, Galileo Galilei (1564–1642) had built at least two telescopes, which he put to use in his study of the universe.

## What **kinds of measurements** do astronomers make with **telescopes**?

Astronomers take lots of carefully planned pictures with telescopes, using a wide variety of telescopes and detectors. These images can then be used to conduct a wide variety of measurements. Aside from examining the images themselves and looking at the shapes and sizes of objects in the universe, some of the most common types of more sophisticated analytical methods include astrometry, photometry, spectroscopy, and interferometry.

## What are the **two main kinds** of **telescopes**?

There are two main types of telescopes: a refractor, which uses lenses to collect light, and a reflector, which uses mirrors. The first telescopes were refractors. Today, almost all of the telescopes being built are reflectors. This is mainly because

229

The earliest astronomers had only their eyes with which to observe space. When astronomers like Galileo, Huygens, and Newton first began to use telescopes, they would meticulously draw their observations onto paper. As technology progressed, new methods to record images and data were developed. Beginning in the nineteenth century, photographic plates became the main recording media of astronomical data for more than 100 years. In the late-twentieth century, photoelectric detectors and computer-based digital cameras replaced photographs. This is the technology by which almost all telescopes today record their data.

large lenses require so much glass that they would quickly sag out of shape from their great weight.

### What is a **Schmidt telescope**?

Invented by the German optician Bernhard Schmidt (1879–1935), this kind of telescope has a primary mirror as its main light-gathering component. This mirror is specially shaped so it can look at a very wide area of the sky at once. Like a "fish-eye" lens on a camera, however, the resulting image is distorted. Thus, a special, thin lens is placed in front of the mirror, which corrects the distortion. This Schmidt design, which uses both refraction and reflection of light, is ideal for obtaining wide-angle images of the sky. It is often used in astronomical cameras.

### What is the **world's largest Schmidt telescope**, and what has it been used for?

The largest Schmidt telescope, the 48-inch (122-centimeter) diameter Oschin Telescope, is at Palomar Observatory on Mount Palomar, California. It was used between the years 1952 and 1959 to conduct the Palomar Optical Sky Survey, the first major systematic photographic survey of the entire northern (and part of the southern) sky. Since then, the survey has been updated using digital camera technology. The telescope has also been used to search for distant Kuiper Belt and Oort Cloud objects. The Oschin Telescope was used to discover many of the largest known Kuiper Belt Objects, such as Varuna, Quaoar, and Eris (which is even larger than Pluto), and also Sedna, thought to be the first Oort Cloud object ever discovered.

### What is the **world's largest refractor**?

With a primary lens diameter of 40 inches (102 centimeters), the great refractor at Yerkes Observatory in Wisconsin is the largest refracting telescope in the world. Built in 1897, it is still in use today. All telescopes larger than that—and built after the end of the nineteenth century—have been reflectors.

## What is the **world's largest reflector?**

Modern reflecting telescopes have mirrors up to 355 inches (8.4 meters) in diameter. A number of telescopes have one primary mirror of approximately that size. Examples include the Subaru Telescope and Gemini North telescopes on Mauna Kea, Hawaii, and the Gemini South telescope on Cerro Pachon, Chile.

The Palomar Observatory in San Diego County, California, is home to the largest Schmidt telescope, having a primary mirror 48 inches in diameter. (*iStock*)

The largest telescopes, though, combine many smaller mirrors together, using them to create an optical system that is equivalent to a telescope with a single large mirror. The Keck 1 and Keck 2 Telescopes on Mauna Kea, Hawaii, each have 36 hexagonal mirrors that fit together to create the equivalent of a telescope 394 inches (10 meters) in diameter. The Large Binocular Telescope on Mount Graham, Arizona, has two eight-meter mirrors on a single mount, creating the equivalent of a single telescope 440 inches (11.2 meters) across. The Very Large Telescope on Cerro Paranal, Chile, is actually four separate telescopes, each eight meters across, positioned side-by-side on the same mountain peak. They are designed to work separately, as well as to work together as a single telescope with an equivalent diameter of 630 inches (16 meters) across.

## What are some of the most ambitious **astronomical surveys** that have been conducted by **observatories** on land and in space?

There have been many large surveys of the cosmos since the middle of the twentieth century. Below is a list of some of the more important ones.

### Notable Astronomical Surveys

| Survey Name | Survey Dates | Characteristics |
| --- | --- | --- |
| NGS-POSS | 1948–1958 | Photographic survey from Mt. Palomar, CA |
| IRAS | 1983 | First far-infrared survey of the entire sky |
| COBE-DMR | 1989–1992 | First survey of the entire cosmic microwave background |
| NRAO VLA Sky Survey | 1993–1996 | Radio survey of continuum emission from VLA, NM |
| FIRST | 1993–2003 | High-resolution radio continuum survey from VLA, NM |
| Hubble Deep Fields | 1995, 1998 | Multi-day Hubble Space Telescope image of 2 patches of sky |

The Mauna Kea observatory, perched on top of a 13,800-foot-high dormant volcano on the Big Island, includes the Keck I and II Telescopes. (*iStock*)

| Survey Name | Survey Dates | Characteristics |
| --- | --- | --- |
| HIPASS | 1997–2002 | Radio survey of atomic hydrogen gas from Australia |
| 2 Micron All Sky Survey | 1997–2001 | Near-infrared survey of the entire sky |
| Sloan Digital Sky Survey | 2000–2005 | Digital optical survey from Sacramento Peak, NM |
| Hubble Ultra Deep Field | 2003–2004 | Deepest astronomical image ever obtained |
| Cosmic Evolution Survey | 2004–2007 | Largest contiguous area surveyed with the Hubble Space Telescope |

# PHOTOGRAPHY AND PHOTOMETRY

### Who **pioneered** the use of **photography** in **astronomy**?

British astronomer William Huggins (1824–1910) was one of the first people to use photographs in astronomy. He used photographic plates exposed over a long period of time—minutes or hours—to record images. Huggins also showed how photographic emulsions could be mixed to increase sensitivity to infrared or ultraviolet light.

### What is **photometry**?

Photometry is the astronomical measurement of brightness (also known as flux or intensity) and color. Photometric intensity is measured as the amount of light energy that strikes a certain surface area on Earth over a certain period of time—in

## How does an astronomical digital camera work for taking images?

**A** digital camera used in astronomy today uses the same basic technology as digital cameras available for purchase at any electronics shop. Light that enters the camera is electronically recorded on a pixellated detector called a charge-coupled device (or CCD for short). When the exposure is finished, an electronic system reads out the information stored on the CCD onto a recording device such as a computer memory stick or hard drive.

The difference is that astronomical sources like planets, stars, and galaxies are so far away that they are almost always too faint to study with typical photographic equipment. In astronomical telescopes, special optical components are therefore used to transmit as much light as possible; CCDs especially efficient in detecting light are used; and the entire camera assembly is cryogenically cooled to hundreds of degrees below zero Fahrenheit in a special container called a dewar. These measures help astronomers measure objects millions—even billions—of times fainter than would be possible with ordinary store-bought cameras.

other words, how bright it appears—and is usually measured in units like "ergs per square centimeter per second," or in terms of apparent magnitude.

## How is **photometry** conducted in **modern astronomy**?

In modern astronomy, photometry is generally conducted using photoelectric detectors or charge-coupled devices. Filters are also used to control the exact wavelengths and colors of light that are measured. This gives astronomers a greater ability to analyze the photometric data scientifically.

Most photometry is obtained using filters that produce standard bandpasses. A bandpass is a well-defined range of wavelengths of light; for example, astronomers refer to the "V-band" as the bandpass of light that ranges in wavelength from about 500 nanometers to 600 nanometers, which encompasses blue-green, green, and yellow light. When astronomers worldwide obtain photometry in common bandpasses, their data and scientific results can be compared, contrasted, and analyzed much more effectively than if everyone used very different bandpasses.

## What do astronomers mean by the **"color"** of an **object**?

In astronomy, the "color" of an object is quantified as the ratio of the brightness measured in two different bandpasses of light. For example, the ratio of U-band light and B-band light is called "(U-B)" and is a measure of the color of any object. When the ratio of the shorter-wavelength bandpass to the longer-wavelength bandpass is higher, then we say that the object is "bluer" in that color; when the ratio is lower, then we say the object is "redder" in that color.

Usually photometry of astronomical objects like stars and galaxies is obtained in more than one bandpass, and from them numerous colors are determined. It is not unusual, for example, to get a galaxy's colors in (U-B), (B-V), (V-R) and (R-K), and to combine all that color information to deduce important properties of that galaxy.

### What do the **colors** of a **distant object reveal** about its properties?

The colors of any object that is warmer than its surroundings are the most important indication of that object's temperature. In a star, for example, stars that are bluer in (U-B) and (B-V) are almost always hotter at their surface than stars that are redder in those colors. These colors help astronomers determine the spectral types of stars: "O" stars have the hottest photospheres, "B" stars are the next hottest, then "A," "F," "G," and "K" stars, and "M" stars are the coolest.

For objects like galaxies and star clusters, which are collections of large numbers of stars, measuring their colors can help astronomers determine how much of the light is coming from what kinds of stars. If the colors of a distant galaxy are bluer, then there are probably more hot stars in that galaxy; and if the colors are redder, then there are probably fewer hot stars.

# SPECTROSCOPY

### What is **spectroscopy**?

Spectroscopy is the process of breaking down light from a source into its component colors to examine the properties of the light source. The detailed pattern of colors that is produced by a light source is called a spectrum. Spectroscopy is like photometry, but in much greater detail. Instead of having relatively large bandpasses, spectroscopy is like photometry with bandpasses just a few nanometers wide, or a few tenths of a nanometer wide, or sometimes even much smaller than that.

A spectrum can be much more complicated than the rainbow of violet (purple), indigo, blue, green, yellow, orange, and red colors with which we are most familiar. When atoms and molecules interact with the light emitted from a source, they can change the spectrum significantly, adding or reducing light in certain colors and patterns. These changes make it possible for astronomers to deduce many of the properties of the light source, and of the intervening material between Earth and the source. Thus, spectroscopy is one of the most important data-analyzing methods astronomers use to learn about the universe.

## Who **pioneered** the use of **spectroscopy in astronomy**?

German physicist Gustav Robert Kirchoff (1824–1887), working with chemist Robsert Bunsen (1811–1899), best known for his Bunsen burner, helped describe how spectroscopy could be used to identify elements. Each kind of atom or molecule interacts with light to produce its own distinct pattern of colors, much the same way that each kind of item in a supermarket can be identified by its own unique bar code. Kirchoff showed that, if light shines through gaseous matter, then the atoms and molecules in that gas would absorb light if the gas were relatively cool, and emit light if the gas were quite hot. Spectroscopic measurements of distant light sources would thus reveal the patterns of dark "absorption lines" and bright "emission lines" produced by the gas. This would in turn reveal the kinds of atoms and molecules in the gas, as well as their environmental conditions. Kirchoff's laws of spectroscopy form the foundation of the analysis of light from distant objects.

Kirchoff measured and studied the spectra of a large number of elements and compounds in his laboratory. He also studied the spectra of stars. Building on Kirchoff's work, British astronomer William Huggins (1824–1910) used photographic technology to record the spectra from very faint and distant stars, opening new avenues of astronomical study. Huggins is known today as the "father" of stellar spectroscopy.

## How is **spectroscopy** used in **modern astronomy**?

Devices called spectrographs are used in conjunction with telescopes and detectors to conduct spectroscopy of objects in the universe. Typically, a modern spectrograph takes light collected through a telescope, usually through a narrow aperture. The incoming light is collimated—made parallel—through a special lens. Then this collimated light bounces through a prism or off a diffraction grating to separate the light into its component colors. The image of the separated light—the spectrum—is then recorded, either photographically or digitally, using a sensitive camera. Once recorded, the spectrum can be analyzed for whatever information it holds about the object that produced it.

## What can we **learn** about **objects** using **spectroscopy**?

When atoms and molecules emit or absorb light, they do so at specific wavelengths of light. When we look at the spectrum of an object, we can deduce what different

**A** major scientific triumph in the early history of astronomical spectroscopy was the discovery of the element helium. Since helium is lighter than air, it leaves Earth's atmosphere unless it is carefully contained. And because of its atomic structure, it is an inert gas and almost never participates in chemical reactions here on Earth. When astronomers first used spectroscopy to study the Sun, though, there were features in the solar spectrum that had never been observed in spectra of terrestrial matter. Scientists realized that a new element had been discovered, and named this new element helium after Helios, the ancient Greek name for the Sun. Eventually, we learned how to gather and use helium here on Earth. Today, we also know that helium is the second most abundant element in the universe, comprising one-quarter of all the atomic mass in the cosmos.

kinds of atoms or molecules are in that object, and the physical conditions and environments that those atoms or molecules are in. Careful spectroscopic study can help us learn about such characteristics as composition, density, temperature, magnetic field strength, and structure. Furthermore, by measuring the Doppler shift of the emission and absorption features in the spectra, we can also deduce how the object is moving; how the different components of the object are moving compared with one another; and, in the case of distant galaxies or quasars, their cosmological redshift can tell us how far away they are and how old they are.

# INTERFEROMETRY

### What is **interferometry**?

Interferometry is a technique of using more than one beam of light at a time to produce images and spectra of especially high resolution or detail. It has many uses, such as measuring the dimensions of very distant objects or the tiny wobble in stellar motions caused by extrasolar planets.

### How does **interferometry work**?

The basic idea behind interferometry is that light travels in waves, and the light waves from one object (or one part of an object) can "interfere," or interact, with the light waves from another object (or another part of that same object). Imagine dropping two pebbles a small distance apart into a pond. The waves made by each pebble interfere with one another, causing wavy ripples of different sizes and shapes. In much the same way, when light waves interfere, they produce similar

> ## What astronomical interferometry system creates the world's largest telescopes?
>
> The most developed astronomical interferometry technologies are currently used in radio astronomy. Perhaps the most impressive example is Very Long Baseline Interferometry (VLBI), which is a method of observation where a number of radio telescopes separated by hundreds or thousands of miles observe the same object simultaneously. The data from the separate telescopes are combined using interferometry to create a single image with the resolution of the widest separation between telescopes—that is, thousands of miles across.
>
> Two of the leading VLBI projects today are the European VLBI Network and the American Very Long Baseline Array (VLBA). Each of those VLBI systems can generate radio-wavelength images that are even sharper than visible-light images taken with the Hubble Space Telescope.

patterns of light, dark, and color. By measuring and studying those interference patterns, astronomers can reconstruct images and deduce other information about the light sources that produced the patterns in the first place, often with much greater detail than would be possible by simply taking an image straight-on.

## How can **interferometry** be used to obtain very **detailed images**?

The resolution of an image depends directly on the size of the telescope used to obtain that image. If, however, the light collected by two telescopes far apart from one another is carefully combined and analyzed using interferometry, the resolution of the resulting image can be as high as if it were obtained using a single telescope as large as the distance between the telescopes. If enough telescopes are used in a row, or in a carefully arranged pattern, it is possible to obtain images hundreds, thousands, or even millions of times more detailed than can be achieved by a single telescope alone.

# RADIO TELESCOPES

## How does a **radio telescope** work?

A radio telescope works very much like the antenna on your car radio. Any long piece of metal can "pick up" radio waves moving past it, and any sheet or scaffolding made of metal can reflect radio waves. Radio telescopes are giant antennae that are specially constructed to reflect radio waves and focus them to a single point. At that point, those waves can be detected, amplified, and interpreted into images or spectra, just like visible light. Since radio waves are millions or even billions of times longer than

237

A line of radio telescopes in Mexico scan outer space for radio waves. (*iStock*)

visible light waves, radio telescopes are generally very large, or consist of large arrays of telescopes that use interferometry to create more detailed images.

### What is the world's **largest radio telescope dish**?

The Arecibo Observatory in Arecibo, Puerto Rico, is operated jointly by the United States National Science Foundation and Cornell University in Ithaca, New York. The Arecibo radio telescope disk is a breathtaking sight. Nestled between hills, on top of a natural valley in the land, it is 1,000 feet (305 meters) in diameter and covers an area of more than 25 football fields. Its Gregorian reflector system is at the focal point of the radio disk, weighs 75 tons, and hangs 450 feet (137 meters) in the air; it is attached to a much larger, 600-ton observing platform, which also hangs there in midair.

Arecibo is by far the world's largest radio dish, and it is the most sensitive radio telescope in the world since its completion in 1963. It has stayed current with regular upgrades to its instrumentation and equipment, and is used day and night for scientific observations and, occasionally, communications with spacecraft far out in the solar system.

### What is the world's **largest steerable radio telescope dish**?

The Robert C. Byrd Green Bank Telescope (GBT) is the world's largest fully steerable radio telescope. It is located at the U.S. National Radio Astronomy Observatory's Green Bank site in Pocahontas County, West Virginia. Another large radio tele-

## Who pioneered radio astronomy?

**A**merican radio engineer Karl Jansky (1905–1950) constructed the first radio telescope and founded the field radio astronomy almost by accident. An employee of Bell Laboratories in New Jersey, Jansky was assigned the task of locating the source of radio interference that was disrupting radiocalls across the Atlantic Ocean. Jansky constructed a radio antenna from wood and brass to detect radio signals at a specific frequency. He found signals coming from three sources: two were thunderstorms, but the third was a mystery that produced a steady hiss. Jansky eventually realized that the signal was being produced by interstellar gas and dust in the Milky Way galaxy. He also observed that the signal was strongest in the direction of the constellation Sagittarius, where we now know the center of the galaxy is located.

In 1932 Jansky's discovery of radio waves from space was announced. The news inspired another American radio engineer, Grote Reber (1911–2002), who proceeded to build his own radio telescope in 1937. For the next decade, Reber studied the radio waves coming from space, creating a map of the radio signals coming from our galaxy. His work showed that most of the radio waves in our galaxy are produced not by stars, but by clouds of hydrogen-rich interstellar gas. Reber's findings, published under the title "Cosmic Static" in *The Astrophysical Journal,* paved the way for a great boom in radio astronomy following the end of World War II in 1945.

scope at Green Bank, which was a slightly smaller telescope than the current one, collapsed in 1988 after 25 years of operation.

The current GBT weighs more than 16 million pounds (7,500 tons) and has a collecting area nearly twice the size of a football field; it is slightly off-axis, and is not exactly round at 110 meters long and 100 meters across. The focal point is at the end of an arm that reaches over the dish from one side. The telescope is mounted on a track 210 feet (64 meters) in diameter that is level to within a few thousandths of an inch. The track allows the telescope to view the entire sky in any direction. Furthermore, each of the 2,004 panels that make up its surface are mounted on motor-driven pistons. This way, the shape of the surface can be carefully adjusted to make very precise observations.

## What is the **VLA radio telescope facility**?

The Very Large Array (VLA) is widely considered to be the world's premier astronomical radio observatory. It consists of 27 radio antennae, each 82 feet (25 meters) across and weighing 230 tons, arranged in a Y-shaped configuration on a high desert plateau near Socorro, New Mexico. The data from all the antennae are combined using interferometry to give the resolution of a single antenna up to 22 miles

The Canberra Deep Dish Communications Complex is one of three global sites comprising NASA's Deep Space Network of radio communications used for space missions. (*NASA*)

(36 kilometers) across, with the sensitivity of a single radio telescope dish 422 feet (130 meters) across.

The VLA is used every day and night to measure and study in great detail distant radio sources, such as pulsars, quasars, and black holes. Each of the 27 antennae is more than seven stories high, and the telescope site is surrounded by beautiful scenery. The VLA has often inspired the creative imaginations of television producers and movie-makers, who have used the facility as a high-technology backdrop for numerous science and science-fiction shows and movies.

# MICROWAVE TELESCOPES

### How does a **microwave telescope** work?

Microwave radiation spans a range of wavelengths that can be produced by very cold astronomical sources, or by warm sources like protoplanetary disks and clouds of interstellar molecules. Microwave telescopes must be able to act somewhat like infrared telescopes and somewhat like radio telescopes. They therefore are built and operated using a fascinating blend of technologies. Depending on the scientific goals, they can be put in space, in high-altitude balloons, or on the ground at mountaintop observatories. Some of the detectors used include bolometers and heterodyne receivers.

### Who **pioneered microwave astronomy**?

Early radio astronomers worked to create radio telescopes in such a way that they could be used to detect microwaves. It turned out, though, that some equipment

## What is the Deep Space Network?

The NASA Deep Space Network (DSN) is an international network of radio antennae that provides communications between Earth-bound scientists and interplanetary spacecraft missions. The DSN has three facilities around the world, so that communications can occur around the clock: one is near Canberra, Australia, one near Madrid, Spain, and one at the Goldstone Apple Valley facility in the Mojave Desert in southern California. The DSN is currently the largest and most sensitive scientific telecommunications system in operation. It also supports some Earth-orbiting missions and is sometimes used for radio astronomy observations of the solar system and beyond.

created for the purpose of wireless communications was the most effective early microwave telescope. In the 1960s, astronomers Arno Penzias (1933–) and Robert Wilson (1936–) used a sensitive microwave antenna built at Bell Laboratories in Murray Hill, New Jersey, to study the microwave radiation from astronomical sources. They discovered the cosmic microwave background, the key evidence that confirmed the Big Bang theory of the origin of the universe.

## What does a **ground-based microwave telescope** look like, and how does it work?

Ground-based microwave telescopes that detect so-called "sub-millimeter" radiation generally look like small radio telescope dishes. They are usually much larger than visible light telescopes, however, and are very carefully constructed and have very sensitive equipment. Some examples are the Submillimeter Telescope (SMT) at the Mount Graham International Observatory in southeastern Arizona; the Swedish-ESO Submillimeter Telescope (SEST) at the European Southern Observatory at La Silla, Chile; and the Caltech Submillimeter Observatory (CSO) on Mauna Kea in Hawaii.

Cold and dry sites are the best places to put microwave telescopes, and the coldest and driest place on Earth is the South Pole. At the Center for Astrophysical Research in Antarctica (CARA), there are two microwave telescopes: the Antarctic Submillimeter Telescope and Remote Observatory (AST/RO) and the Cosmic Background Radiation Anisotropy (COBRA) experiment. These telescopes look somewhat different from microwave telescopes at other sites in order to adapt to the harsh environmental conditions at the pole.

## What was the **COBE satellite**?

The Cosmic Background Explorer (COBE) was used to make detailed microwave maps of the universe. COBE was launched on November 18, 1989, and had three scientific instruments on board: the Far Infrared Absolute Spectrophotometer

(FIRAS), the Differential Microwave Radiometer (DMR), and the Diffuse Infrared Background Explorer (DIRBE). Data gathered by COBE confirmed the presence of the cosmic microwave background that was evidence of the Big Bang. COBE measured the temperature of the background to be 2.7 degrees Kelvin (almost absolute zero) and also showed that the imperfections in the background indicated the origins of structure in the universe.

## What is the **WMAP satellite**?

The Wilkinson Microwave Anisotropy Probe (WMAP) is a microwave space telescope designed to measure the tiny variations, or anisotropies, in temperature that exist in the cosmic microwave background radiation. By measuring how strong the anisotropies are, how many there are, and how large they are, astronomers can trace the evolution of the early universe and deduce fundamental properties of the universe as a whole. The probe is named for the American astrophysicist David Wilkinson (1935–2002), who among his many scientific achievements was a pioneer in the measurement and study of the cosmic microwave background.

WMAP has made a tremendous impact on astronomers' understanding of the cosmos. Perhaps its most important results are the confirmation that the universe has a so-called flat geometry, and that more than 70 percent of the contents of the universe is comprised of a mysterious "dark energy."

# SOLAR TELESCOPES

## How does a **ground-based solar telescope** work?

The optics and detectors of a solar telescope are similar to telescopes that are used primarily at night. What is different is that solar telescopes must be constructed to account for the intense brightness and heat they experience. One way to keep the telescope components and instruments cooler is to direct the light into an underground chamber first. Another way is to maintain a vacuum around the telescope, because in a vacuum no air molecules are present to absorb and transfer the heat. Whereas most nighttime telescopes are designed to have as large a primary mirror as possible, but otherwise be light and maneuverable, the primary mirrors of solar telescopes are not particularly large. The equipment and building structures associated with them, though, are often huge.

### How does a **space-based solar telescope** work?

Astronomers use orbiting solar telescopes to observe the Sun using wavelengths of light that do not easily penetrate Earth's atmosphere, or to observe subatomic particles like solar wind or coronal mass ejections that are blocked by Earth's magnetic field. Like ground-based solar telescopes, these solar space telescopes are similar to other space telescopes; they are adjusted to accommodate the strong flux of both radiation and particles from their target.

### What are some **well-known space-based solar telescopes**?

Some well-known examples of solar space telescopes are the Solar Maximum Mission (SMM), which operated from February 1980 to November 1989; the Solar and Heliospheric Observatory (SOHO), launched on December 2, 1996; and the Transition Region and Coronal Explorer (TRACE), launched on April 2, 1998.

# SPECIAL TELESCOPES

### What **telescope uses ice** to study the universe?

When neutrinos penetrate matter the collision causes a brief flash of bluish light called Cherenkov radiation. If such a flash occurs in a block of ice that is free of air bubbles or other impurities, the Cherenkov light can be detected by sensitive photosensors. Astrophysicists have taken advantage of this unusual property of ice to build the world's largest neutrino telescope. The Antarctic Muon and Neutrino Detector Array (AMANDA) project consists of 19 long chains of photodetectors embedded more than a mile deep in the Antarctic ice at the South Pole. AMANDA is part of an even larger project called IceCube, an international scientific project that will eventually suspend thousands of photodetectors throughout a cubic kilometer of Antarctic ice.

### What **telescopes** can be used to observe **cosmic rays**?

Cosmic rays are so energetic that they pass through just about any obstacle, including Earth itself. Occasionally, though, they will collide with matter in Earth's atmosphere; the interaction causes a cascade of electromagnetic radiation to spew forth, which is called a Cherenkov shower. To study these powerful cosmic rays, astronomers have built "atmospheric Cherenkov detectors." By analyzing the showers, scientists can deduce important properties about the cosmic rays that created them.

### What are some well-known **atmospheric Cherenkov systems**?

Some well-known atmospheric Cherenkov systems include VERITAS (Very Energetic Radiation Imaging Telescope Array System), located at the Fred Lawrence Whipple Observatory on Mount Hopkins, Arizona; and the Solar Tower Atmospheric Cerenkov Effect Experiment (STACEE) at the National Solar Thermal Test Facility (NSTTF) of Sandia National Laboratories in New Mexico.

### What **observatory** can be used to **observe** the **buckling of space** itself?

When a powerful, explosive event occurs in the universe, such as a supernova explosion or the collision of two black holes, space itself is affected. Astrophysicists have created the Laser Interferometer Gravitational-Wave Observatory (LIGO) to try to detect the gravitational waves produced in such events. There are two facilities—one in the state of Louisiana, another in the state of Washington—with super-sensitive laser interferometric systems situated deep underground. So far, there has been no detection.

# TERRESTRIAL OBSERVATORIES

### What is an **observatory**?

An observatory is a facility where astronomical observations can take place. They can consist of just one telescope, but often they have many telescopes. Modern observatories sometimes do not even have a telescope at the location; instead, they are the locations where scientists gather to obtain and analyze data, even if the telescope they are using is far away on Earth or in space.

### How do astronomers decide **where** to **build observatories**?

Astronomers today spend years examining potential observatory sites to find the best places in the world to build and install telescopes. Ideally, a telescope should be at a high altitude, at a site free of air or light pollution, where the atmospheric flow is calm and predictable, where the impact on the ecological environment is relatively small, and where humans, machinery, and equipment can go and be safe and well-maintained.

Since there are a limited number of such sites worldwide, good observatory locations often wind up with many telescopes in the same location. As the world's population has grown by leaps and bounds, and as the scientific requirements for a good site become increasingly demanding, astronomers have had to seek ever more remote sites for creating observatories, such as the Atacama Desert in Chile; the Alta Plana in Peru; remote mountains in Mexico; and ocean archipelagos like the Hawaiian and Canary Islands.

## Where are the world's **largest visible light telescopes**?

The following table lists the largest telescopes in order of their mirror sizes.

### World's Largest Visible Light Telescopes

| Name | Location | Aperture Equivalent Diameter (in meters) | Configuration |
| --- | --- | --- | --- |
| Large Binocular Telescope | Mount Graham, AZ | 11.8 | two circular mirrors |
| Great Canary Telescope | La Palma, Canary Islands | 10.4 | 36 hexagonal mirrors |
| Keck I | Mauna Kea, HI | 10.0 | 36 hexagonal mirrors |
| Keck II | Mauna Kea, HI | 10.0 | 36 hexagonal mirrors |
| Hobby-Eberly Telescope | McDonald Observatory, TX | 9.2 | 91 hexagonal mirrors |
| South African Large Telescope | Sutherland, South Africa | 9.2 | 91 hexagonal mirrors |
| Subaru | Mauna Kea, HI | 8.2 | one circular mirror |
| Antu | Cerro Paranal, Chile | 8.2 | one circular mirror |
| Kueyen | Cerro Paranal, Chile | 8.2 | one circular mirror |
| Melipal | Cerro Paranal, Chile | 8.2 | one circular mirror |
| Yepun | Cerro Paranal, Chile | 8.2 | one circular mirror |
| Gemini North | Mauna Kea, HI | 8.1 | one circular mirror |
| Gemini South | Cerro Pachon, Chile | 8.1 | one circular mirror |
| MMT | Mt. Hopkins, AZ | 6.5 | one circular mirror |
| Walter Baade Telescope | Las Campanas, Chile | 6.5 | one circular mirror |
| Landon Clay Telescope | Las Campanas, Chile | 6.5 | one circular mirror |
| Large Altazimuth Telescope | Nizhny Arkhyz, Russia | 6.0 | one circular mirror |
| Liquid Zenith Telescope | British Columbia, Canada | 6.0 | liquid, points up |
| Hale Telescope | Mt. Palomar, CA | 5.0 | one circular mirror |

The European Southern Observatory (ESO) is a system of astronomical facilities run by a consortium of European nations. It is headquartered in Garching, Germany, but as its name implies, its observing facilities are in the southern hemisphere—specifically, northern Chile. Its main site is the Very Large Telescope (VLT)—a suite of four big telescopes on Cerro Paranal. Its original site, which still is home to many telescopes run by ESO and many of its member nations, is on the mountain La Silla.

## What are the **national observatories** of the **United States**?

The national observatories of the United States, which are funded primarily by the U.S. National Science Foundation, are the National Optical Astronomy Observatory (NOAO) and the National Radio Astronomy Observatory (NRAO). NOAO and NRAO both operate several facilities: NOAO has the National Solar Observatory; Kitt Peak National Observatory in southern Arizona; Cerro Tololo Inter-American Observatory near La Serena, Chile; and the NOAO Gemini Science Center, which is responsible for one telescope on Cerro Pachon, Chile, and another on Mauna Kea, Hawaii. Its main scientific office is in Tucson, Arizona. NRAO has the Very Large Array near Socorro, New Mexico; the Green Bank Radio Observatory at Green Bank, West Virginia; the Very Long Baseline Array, which spans more than 5,000 miles; and the Atacama Large Millimeter Array in the Chajnantor plain of the Atacama Desert in northern Chile. NRAO's main office is in Charlottesville, Virginia.

## What astronomical **observatories** are in **Australia**?

Australia has many well-known radio telescopes, including the Parkes Radio Telescope, which was used to communicate with the Apollo missions to the Moon. The premier astronomical research facility in Australia is the Mount Stromlo and Siding Spring Observatories, which is run by the Australia National University.

## What astronomical **observatories** are in **Africa**?

The best-known astronomical observatory in Africa is the South African Astronomical Observatory (SAAO), with offices near Cape Town. Its telescope site is at Sutherland, in the Karoo region of South Africa, and its largest telescope is the South African Large Telescope (SALT), which was first active in 2005.

## What astronomical **observatories** are in **Asia**?

The best-known telescope facility in Asia is probably the six-meter Bol'shoi Teleskop Azimultal'nyi (BTA, or "Great Azimuthal Telescope"), near Nizhny Arkhyz, Russia. The telescope sits atop Mount Pastukhov, about 90 miles (150 kilometers) south of

## What are some well-known observatories in Britain?

**A**lthough there are no major research telescopes located in the British Isles, there has been a rich history of astronomical research centered in Britain for centuries. The Jodrell Bank Centre for Astrophysics, for example, is still a major radio astronomy research center in Manchester. The Royal Observatory in Greenwich is no longer an astronomical research facility, but it is still significant as the origin for the Prime Meridian and the longitude system that maps Earth's surface and the celestial sphere.

Stavropol, between the Black Sea and the Caspian Sea. It has been operating since 1976, and for a time was the world's largest telescope.

## What astronomical **observatories** are in the **Atlantic Ocean**?

The Canary Islands is the site of the Observatorio del Teide at Tenerife and Observatorio del Roque de los Muchachos at La Palma. They are currently home to telescopes and other instruments belonging to 60 scientific institutions from 17 different countries. These observing facilities, together with others, make up what is currently called the "European Northern Observatory" (ENO).

## What astronomical **observatories** are in the **Pacific Ocean**?

A number of important astronomical observatories are located in the Hawaiian Islands, including what is probably the best ground-based telescope site in the world: the summit of the extinct volcano Mauna Kea on the Big Island of Hawaii. Some important telescopes there include the National Astronomical Observatory of Japan's Subaru Telescope, the Gemini North Telescope, the James Clerk Maxwell Telescope, the United Kingdom Infrared Telescope, the Canada-France-Hawaii Telescope, and the twin Keck Telescopes.

## What are some well-known **university-run observatories**?

Among the best known university observatories are the Harvard-Smithsonian Center for Astrophysics; the University of California's Lick Observatory; the California Institute of Technology's Palomar Observatory, and, in collaboration with the University of California, Keck Telescopes; the telescopes on Mauna Kea, Hawaii, and Mees Solar Observatory on Maui, operated in part by the University of Hawaii; and the University of Arizona's Steward Observatory, Arizona Radio Observatory, and Mount Graham International Observatory.

## What are some well-known **private observatories**?

The Smithsonian Astrophysical Observatory, a bureau of the Smithsonian Institution, has been teamed with the Harvard College Observatory for many years to run

247

Many universities around the world operate their own observatories, including the Yerkes Observatory at the University of Chicago. (*iStock*)

the Harvard-Smithsonian Center for Astrophysics, a research facility based in Cambridge, Massachusetts, with some 300 scientists. The Lowell Observatory, based in Flagstaff, Arizona, was founded by Percival Lowell a century ago to search for, among other astronomical phenomena, Planet X; today, it remains a major center for astronomical research. Perhaps the largest and most prestigious private observatory today is the Observatories of the Carnegie Institution of Washington. The wealthy industrialist Andrew Carnegie founded the Carnegie Institution in 1902, and it has been a major force in astronomical research ever since. The Carnegie Observatories currently operate facilities in Pasadena, California, and Las Campanas, Chile, which is home to the twin Magellan Telescopes. Carnegie Institution's Washington offices is also home to the Department of Terrestrial Magnetism, which has had a very strong astronomical presence for nearly a century. Astrophysicists there have pioneered research on dark matter, astrobiology, and exoplanets.

# AIRBORNE AND INFRARED OBSERVATORIES

### How does an **infrared telescope** work?

Infrared radiation can be divided roughly into near-infrared, mid-infrared or thermal-infrared, and far-infrared categories. Most ground-based telescopes on Earth can be used for observations of both visible light and near-infrared radiation; mid-infrared light can be observed from Earth or from space; far-infrared radiation can only be

## What is a virtual observatory?

The term "virtual observatory" was coined about a decade ago to describe a facility or set of facilities, linked by computer networks, that allow astronomers to conduct scientific studies using data previously obtained with actual telescopes. The astronomical community has, over many decades, compiled a vast archive of astronomical data that have been obtained for specific scientific goals. A great deal of the data, however, still has significant information about the universe waiting to be discovered. All the world's major astronomical observatories have agreed to share their archived data with all the scientists of the world. They have cooperated to develop virtual observatory facilities that will allow this data to be explored and analyzed to make new and exciting discoveries that could not be made with any single telescope alone. The International Virtual Observatory Alliance (IVOA), with more than a dozen national member facilities, is hard at work in creating virtual observatory tools that will benefit all astronomers and scientists worldwide.

observed effectively from space. Generally, infrared telescopes look and operate much like visible-light telescopes. However, since infrared radiation is a form of heat, telescopes and cameras used to observe infrared emissions from space work most effectively if they are cryogenically cooled—often using liquid helium, to temperatures less than ten degrees above absolute zero. The digital detectors are also made of different substances, to increase their sensitivity to infrared radiation. Whereas most visible light detectors are made primarily of silicon, near-infrared detectors are often made of germanium, or exotic materials like gallium arsenide, indium antimonide, or a blend of mercury, cadmium, and tellurium called "mer-cad-telluride."

## What are some **examples** of an **infrared telescope**?

Examples of space-based infrared telescopes are the Infrared Astronomical Satellite (IRAS) and the Spitzer Space Telescope. Examples of ground-based infrared telescopes are the Infrared Telescope Facility (IRTF) and the United Kingdon Infrared Telescope (UKIRT), both on Mauna Kea in Hawaii.

## What is an **airborne observatory**?

An airborne observatory is a telescope installed in an airplane that is operated while the plane is in flight.

An image of the Lagoon Nebula as seen with infrared light through the Hubble infrared telescope. (*NASA*)

249

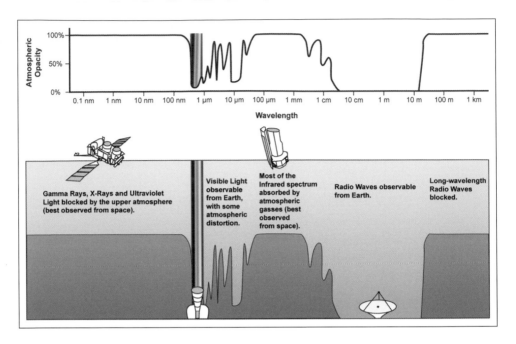

Earth's atmosphere can block various wavelengths of light and other energy. Airborne and space observatories are needed so that this interference can be overcome. (*NASA/IPAC*)

## Why would **astronomers** want to use an **airborne observatory**?

When an airplane flies at an altitude of about 41,000 feet (12,500 meters), it is above 99 percent of Earth's atmospheric water vapor. Since water vapor absorbs incoming infrared radiation, being above the vapor makes it possible to make many infrared observations. At the same time, airborne telescopes are easily accessible for repairs, upgrades, and real-time adjustments, unlike space telescopes. Also, the airplane can be flown to different parts of the world and operated at different places, offering greater flexibility than a terrestrial telescope.

## Are there **disadvantages** to **airborne observatories**?

Yes, there are disadvantages. First, it is more expensive and technically difficult to operate a telescope on a flying airplane than it would be to do so from a fixed location on the ground. Second, an airplane has significant limits in the amount of weight it can carry, so any airborne telescope is likely to be relatively small compared to ground-based telescopes. Hence, airborne observatories are almost exclusively used for only far-infrared observations, where they have the greatest advantage over facilities on the ground.

## What was the **Kuiper Airborne Observatory**?

The Kuiper Airborne Observatory (KAO), named after the Dutch-American astronomer Gerard Kuiper (1905–1973), was operated by NASA, beginning in 1974. It was a modified C-141 Starlifter military cargo aircraft and carried a 36-inch (0.9-

meter) diameter telescope weighing about 6,000 pounds (2,700 kilograms). It was based at the NASA Ames Research Center near Moffat Field, California, and flew about 70 scientific flights per year.

The KAO was retired in 1995, after more than 20 years of scientific service. Among the many discoveries it fostered were planetary rings around Uranus; water vapor in Jupiter's atmosphere; the thin atmospheric layer that sometimes surrounds Pluto; and pioneering studies of comets, asteroids, and the interstellar medium.

## What is **SOFIA**?

The Stratospheric Observatory for Infrared Astronomy (SOFIA) is an airborne infrared observatory that is expected to be the successor to the Kuiper Airborne Observatory (KAO). It is a Boeing 747 airliner that has been modified to carry a 100-inch (2.5-meter) diameter telescope, which will operate at altitudes of about 41,000 feet (12,500 meters). SOFIA is a joint project of NASA and the German Aerospace Center (DLR). It made its first flight on April 26, 2007.

# SPACE TELESCOPES

## Why is it **important** to have **telescopes in space**?

The thick layer of gas that is in our atmosphere blocks most of the electromagnetic radiation that reaches our planet from astronomical sources, including gamma rays, X rays, far-ultraviolet and far-infrared light. Atmospheric disturbances such as wind, rain, and snow, also block the view of space. Thus, space telescopes can take much clearer pictures than can be taken from Earth; they can collect light that simply does not reach the surface of Earth.

## What is the **best-known space telescope**?

The Hubble Space Telescope is named after the American astronomer Edwin Hubble. It was launched on the space shuttle *Discovery* on April 24, 1990, and is operated by NASA and the European Space Agency. Since then, this telescope (known as HST for short) has transformed our understanding of the universe more than any other telescope in history. Both scientifically and socially, this space telescope has been the most influential scientific facility of our generation.

251

The Hubble Space Telescope. (*NASA*)

## What are some **specifications** of the **Hubble Space Telescope**?

The Hubble Space Telescope just barely fits into the cargo bay of the space shuttle. It is about the size of a large school bus, and its cameras and spectrographs are about the size of phone booths. Its total mass is about 13 tons. The primary mirror of HST is 94 inches (2.4 meters) across—that is relatively small compared to modern ground-based telescopes, but by far the largest telescope ever launched into space. While HST is in operation, it has two large solar panels that generate electricity to power the telescope's many systems.

## What is the **orbit** of the **Hubble Space Telescope**?

The Hubble Space Telescope orbits about 240 miles above Earth's surface. It races around our planet once every 90 minutes.

## How was the **Hubble Space Telescope project constructed** and **deployed**?

The initial proposal for an orbiting telescope was made in 1946 by the American astronomer Lyman Spitzer, Jr. (1914–1997). In the early 1970s, as the Apollo program came to an end, NASA accepted an initial proposal for a space telescope. However, the U.S. Congress delayed the project because of the expense. In 1977 the European Space Agency joined the United States as a partner, agreeing to provide 15 percent of the support and equipment needed for the space telescope project in exchange for 15 percent of the telescope's observing time.

The construction of the Hubble Space Telescope took eight years and about 1.5 billion dollars. It was completed in 1985. The launch was delayed after the 1986 space shuttle *Challenger* disaster, which grounded the shuttle fleet for more than two and a half years. The HST was finally deployed by the space shuttle *Discovery* on April 24, 1990.

## What **happened** to the **Hubble Space Telescope** after it was **launched**?

After HST was launched and placed into orbit in 1990, initial tests showed that the telescope had several significant flaws. The solar panels jittered ever so slightly as the telescope orbited, blurring the images it took; worse, the primary mirror had been polished to the wrong shape. The result was an optical effect called spherical aberration in the images, which degraded the image quality by almost 90 percent. This was a tremendous blow to astronomers, who had been eagerly anticipating the clearest pictures of the universe ever taken up to that time.

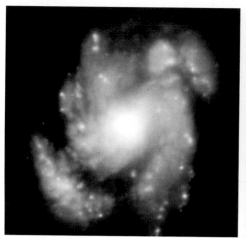

When the Hubble's mirror was replaced there was a profound improvement in the quality of the images it returned to Earth observers. (*NASA*)

## How was the **Hubble Space Telescope repaired**?

It took months to characterize and measure exactly what was wrong with the telescope. Once that was done, new solar panels were constructed, and special optical equipment was built to compensate for the aberrant optics. Principal among these plans were the Corrective Optics Space Telescope Axial Replacement (COSTAR), a set of three coin-sized mirrors that would bring light from the primary mirror into proper focus, and a new camera called the WFPC–2 (the second Wide Field/Planetary Camera).

The necessary servicing to the HST was conducted in December 1993 by four astronauts on the space shuttle *Endeavour*. They caught up with the Hubble two days after launch, and used the shuttle's robotic arm to bring it into the shuttle's cargo bay. For nearly a week, they conducted the necessary repairs and maintenance on the telescope. The first images that came down from the repaired telescope were transmitted early in the morning of December 18, 1993.

## What are considered to be **NASA's four Great Observatories** in space?

NASA's four Great Observatories in space are the Compton Gamma Ray Observatory (CGRO), the Hubble Space Telescope, the Chandra X-ray Observatory, and the Spitzer Space Telescope.

## What are some of the most **significant discoveries** made with the **Hubble Space Telescope**?

There is only enough room in this book to mention a very few of the incredible cosmic discoveries that HST has enabled astronomers to make. Using Hubble, astronomers have taken the deepest-ever image of the universe; found the most distant galaxies in the universe (13 billion light-years away); measured the cosmic expansion rate; helped confirm the existence of "dark energy" and the accelerating expanion of the universe; discovered two tiny new moons orbiting Pluto; and proved that there are supermassive black holes in most spiral and elliptical galaxies.

### What will be the **successor** to the **Hubble Space Telescope**?

If all of the servicing and updating of the Hubble Space Telescope is completed as planned, then it is scheduled to continue operating as a cutting-edge scientific instrument until at least the year 2015. Around that time, the successor to the Hubble Space Telescope will be launched. Already under development, the James Webb Space Telescope will have a folding, segmented primary mirror nearly 10 times the size of the HST's.

# INFRARED SPACE TELESCOPES

### What was the **IRAS telescope**?

The Infrared Astronomical Satellite, or IRAS, was launched in 1983 by an international science consortium that included the United States, Britain, and the Netherlands. The main telescope on IRAS was a 23-inch (58-centimeter) reflector telescope. The infrared light collected by this telescope was measured and recorded by 64 semiconductor panels, and sent down to Earth by radio signals. To keep the scientific instruments cool, they were surrounded by a large insulated flask of liquid helium. The IRAS instruments operated until the liquid helium, which lasted about seven months, boiled away into space.

### What did the **IRAS telescope observe**?

Over its seven-month span, IRAS surveyed the sky twice over at mid-infrared and far-infrared wavelengths. These all-sky surveys were the first of their kind and opened many new avenues of astronomical study. Among its most signifi-

The Spitzer Space Telescope is one of the current Great Observatories gathering data from outer space. (*NASA/JPL-Caltech*)

cant scientific achievements were the discovery of a new class of very bright galaxies that emit the vast majority of their light at infrared wavelengths; the identification of stellar nurseries, heavily obscured by dusty gas clouds, where new stars are being born; and the discovery and mapping of so-called "infrared cirrus," vast clouds of very sparse interstellar gas and dust that glow faintly at far-infrared wavelengths.

## What was the **ISO telescope**?

The Infrared Space Observatory (ISO) was a successor to the Infrared Astronomical Satellite (IRAS). ISO was launched in November 1995, and was activated on November 28, 1995. It was about the same size as IRAS, and was constructed in about the same way. It had much more sensitive infrared instruments, however, including two infrared spectrographs. Unlike IRAS, which surveyed as much of the sky as it could, ISO was targeted at specific astronomical objects and regions for detailed studies. It helped pave the way for the current-generation infrared space telescope, the Spitzer.

## What is the **Spitzer Space Telescope**?

The Spitzer Space Telescope (SST) is the largest and most sophisticated infrared space telescope ever launched. Like the Infrared Space Observatory (ISO), SST is designed for targeted observations, rather than all-sky surveys. It is, however, much larger than ISO, and it has much better image-taking and spectroscopic resolution and sensitivity, as well. The Spitzer Telescope was launched on August 25, 2003, and its first images were released to the public on December 18 of that year.

255

## What did the **Spitzer Space Telescope discover**?

Among the many discoveries already made with Spitzer are brand-new populations of brown dwarf stars, a number of exoplanets and protoplanetary disks around young stars, distant galaxies and quasars obscured by huge amounts of interstellar dust, and the most detailed map of the center of the Milky Way galaxy ever made.

# X–RAY SPACE TELESCOPES

## How does an **X-ray telescope work**?

X rays are so powerful that they tend to pierce right through typical telescope mirrors if they strike them head-on. Therefore, X-ray telescopes use nested layers of "grazing-incidence mirrors" that reflect X rays along very shallow angles. The need for grazing incidence optics makes X-ray telescopes very challenging to design. (Compared to optical telescopes, they often look like they are pointed backward!) Furthermore, X rays do not go through the atmosphere well, so all X-ray telescopes must be space telescopes. The scientific reward, however, is well worth the difficulty of building them. X-ray telescopes afford astronomers an opportunity to study directly some of the most energetic phenomena in the universe, like novae, supernovae, pulsars, and black holes.

## What were the **first X-ray telescopes**?

In 1962 a scientific team led by Italian-American physicist Ricardo Giacconi (1931–) and his colleagues launched an X-ray telescope into space aboard an Aerobee rocket. The flight lasted only a few minutes, but during the time it was above the absorptive layers of the atmosphere, the telescope detected the first X rays from interstellar space, including a strong source coming from the direction of the constellation Scorpius. Subsequent flights during the 1960s detected X rays from many other sources, including one toward the constellation Cygnus and another toward Taurus (the Crab Nebula).

## What was the first **X-ray telescope satellite**?

The first satellite designed specifically for X-ray astronomy was *Uhuru*, which means "peace" in Swahili. *Uhuru* was launched in 1970, and during its lifetime produced the first X-ray map of the sky.

## What were the **HEAO missions**?

The three High-Energy Astrophysics Observatory (HEAO) missions were a series of telescopes launched by NASA to study X rays, gamma rays, and cosmic rays. During its year and a half of operation, HEAO-1 provided continuous monitoring of many astronomical X-ray sources, such as quasars and pulsars. HEAO-2, also called the Einstein Observatory, operated from November 1978 to April 1981, and took the highest resolution X-ray images of the sky up to that time. HEAO-3 was launched in 1979 and focused on gamma rays and cosmic rays. Einstein helped pave the way for Chandra, NASA's flagship X-ray observatory mission, while HEAO-3 led to the Compton Gamma Ray Observatory.

A researcher at the Marshall Space and Flight Center in Huntsville, Alabama, conducts studies to develop technology for an X-ray telescope. (*NASA/Dennis Keim*)

## What was the **ROSAT mission**?

The Roentgen Satellite (ROSAT) mission was the next generation X-ray space telescope after the HEAO missions. It was led by a German scientific team, and is named after the German physicist Wilhelm Conrad Roentgen (1845–1923), who discovered X rays.

## What is the **XMM-Newton mission**?

The X-ray Maximum Mission (XMM-Newton) satellite is the major European X-ray observatory class space telescope, and the largest orbiting scientific payload ever built in Europe. It was launched on December 10, 1999, and is operated by the European Space Agency. In addition to its main X-ray telescope—the most sensitive ever built—it also carries a smaller monitor telescope that works in ultraviolet and visible light. That co-aligned telescope points at the same location in space as the X-ray telescope and allows astronomers to locate the X-ray sources immediately in a detailed visible light image.

## What are some **discoveries** made by the **XMM-Newton mission**?

A few of the many discoveries made with XMM-Newton are the direct detection of matter falling onto black holes; detailed studies of supernovae and other stellar explosions; new observations of white dwarf stars, neutron stars, and quasi-stellar objects; and pioneering observations of gamma-ray bursts.

## What is the **Chandra X-ray Observatory**?

The Chandra X-ray Observatory is the most capable X-ray telescope mission ever launched by NASA. It was launched on July 23, 1999, aboard the space shuttle *Columbia*. It orbits Earth on a highly elliptical trajectory, coming as close as 6,200 miles (10,000 kilometers) and going as far as 87,000 miles (140,000 kilometers) from Earth's surface. Although challenging to operate as a result, this extreme orbit allows scientists to make observations of many different kinds that would not be possible at only one orbital altitude. Chandra obtains X-ray images of astronomical objects with much greater resolution than any other X-ray telescope ever flown; thus, it has returned detailed images of complex, highly energetic astronomical systems, such as supernova remnants, supermassive black hole systems (including Sagittarius A*, the one at the center of our Milky Way galaxy), shock waves from exploding stars, the X-ray shadow of Saturn's moon Titan, and multi-million degree gas in dense clusters of galaxies.

# ULTRAVIOLET SPACE TELESCOPES

## How does an **ultraviolet telescope work**?

Ultraviolet telescopes, like X-ray telescopes, need to be above Earth's atmosphere in order to be effective. The mirror technology is about the same as that of visible light telescopes; the detectors, though, have to be specially built to be sensitive to ultraviolet light. For example, they use specially constructed charge-coupled devices

(CCDs) and Multi-Anode Microchannel Arrays (MAMAs). The largest ultraviolet space telescope ever launched is the Hubble Space Telescope.

## What were the **first ultraviolet telescopes** to be deployed in space?

The first ultraviolet telescopes included eight Orbiting Solar Observatories (OSO), which were launched between 1962 and 1975. The OSO measured ultraviolet radiation from the Sun. The data collected from these telescopes provided scientists with a much more complete picture of the solar corona. A series of Orbiting Astronomical Observatories (OAO) were also deployed, and were used to study ultraviolet emissions from objects other than the Sun, including thousands of stars, a comet, and a number of galaxies. From 1972 to 1980, OAO Copernicus collected ultraviolet data on a number of stars, as well as on the temperature, composition, and structure of the interstellar medium.

## What was the **IUE satellite**?

The International Ultraviolet Explorer (IUE) was launched in 1978, and it obtained its first ultraviolet spectrum of an astronomical object on January 26, 1978. Although its last scientific spectrum was obtained on September 30, 1996, it remains in orbit today. IUE obtained more than 104,000 observations and gave astronomers the first accurate information about the ultraviolet properties of planets, stars, and galaxies—data that are still being used today to interpret more current observations.

## What was the **EUVE satellite**?

The Extreme Ultraviolet Explorer (EUVE) mission was the first telescope dedicated to observing the universe in the shortest wavelengths of ultraviolet light. This kind of ultraviolet radiation is almost as energetic as X-ray radiation, so EUVE was built using a combination of ultraviolet and X-ray telescope technology. It was launched on June 7, 1992, and operated until January 31, 2001. Among its scientific accomplishments were an all-sky survey catalog of 801 objects; the first extreme ultraviolet detections of objects outside the Milky Way galaxy; the detection of extreme UV photospheric emission from stars; and observation of particular behaviors of unusual objects such as quasi-periodic oscillations in a dwarf nova.

## What was the **FUSE satellite**?

The Far Ultraviolet Spectroscopic Explorer (FUSE) was launched on June 24, 1999. It was the scientific successor to the International Ultraviolet Explorer (IUE) and employed many new technologies, including the use of four mirror segments rather than a single primary mirror. FUSE was designed to study, among many other things, the distribution and composition of matter early in the universe; the dispersion of chemical elements throughout galaxies; and the properties of interstellar gas clouds out of which stars and solar systems form. After a successful three-year primary mission, FUSE embarked on a longer secondary mission that achieved many more scientific milestones, including discoveries about deuterium in the distant universe and observations of hundreds of stars in the Magellanic Clouds. FUSE was decommissioned on October 18, 2007.

## What is the GALEX mission?

The Galaxy Evolution Explorer (GALEX) mission is an orbiting ultraviolet space telescope that was launched on April 28, 2003. A Pegasus rocket placed GALEX into a nearly circular orbit at an altitude of 432 miles (697 kilometers). Its primary mission has been to take ultraviolet images and photometric measurements of more than 100,000 galaxies, stars, and other astronomical objects. GALEX was designed to be particularly sensitive to very hot stars—usually either very young and luminous main sequence stars, or hot white dwarfs—and thus has made very important discoveries about the star formation histories and processes in both nearby and distant galaxies.

# GAMMA–RAY SPACE TELESCOPES

## How does a gamma-ray telescope work?

Gamma rays are the most energetic form of electromagnetic radiation, and as a result they penetrate any kind of mirror material, even at a grazing incidence angle. Technologies incorporated in gamma-ray telescopes are thus very different from any other kinds of telescopes. They include plastic, gas, and crystal scintillation detectors; coded aperture masks and arrays; and spark chambers and silicon strip detectors.

## What were the first gamma-ray telescopes?

Early space telescopes detected some gamma rays, even though they were very inefficient and imprecise. Explorer XI in 1961 detected a tiny gamma-ray flux. The third of the Orbiting Solar Observatory satellites (OSO-3) was used to detect astronomical gamma rays in 1967; and SAS-2 was also a valuable early gamma-ray observatory that was launched in 1972.

## What was the COS-B satellite?

COS-B was an X-ray and gamma-ray telescope that was operated by the European Space Agency from August 9, 1975, to April 25, 1982. Important scientific results achieved by COS-B included the first gamma-ray map of the Milky Way galaxy; observations of the Cygnus X-3 pulsar; and a catalog of 25 powerful gamma-ray sources.

## What was the Compton Gamma Ray Observatory?

The Compton Gamma Ray Observatory (CGRO) was the high-energy astrophysics mission that was one of NASA's four Great Observatories in space. CGRO was launched on April 5, 1991, and started working perfectly, conducting more than nine years of pioneering scientific observations.

## What accidental discovery led to the blossoming of gamma-ray astronomy?

In the 1960s and 1970s, the United States government launched satellites that orbited Earth to monitor the nuclear test ban treaty that was in place between the United States and the Soviet Union. The detectors on the satellites were designed to see bursts of gamma rays coming from Earth's surface that would indicate a nuclear explosion. Surprisingly, these satellites detected a gamma-ray burst every few days, but none of them were coming from Earth. Scientists realized that these gamma-ray satellites had discovered a new kind of astronomical phenomenon. After the data from this military mission was publicly announced, the gamma-ray burst phenomenon energized this area of astronomy, and has helped to drive research and telescope design in the field.

The CGRO had four major scientific instruments, all of which produced significant scientific discoveries in high-energy X-ray and gamma-ray astronomy. The Burst and Transient Source Experiment (BATSE) monitored the entire sky for stellar explosions and gamma-ray bursts, and helped prove that gamma-ray bursts are hugely powerful explosions that usually occur in galaxies other than the Milky Way. The Compton Telescope (COMPTEL) imaged nearly a tenth of the sky at a time, and the Oriented Scintillation Spectrometer Experiment (OSSE) made more detailed observations in smaller areas of the sky. Together, they made the most sensitive and detailed gamma-ray maps and studies ever of the Sun, our galaxy, and the entire sky. (OSSE even found evidence of streams of antimatter in the Milky Way!) Finally, the Energetic Gamma Ray Experiment Telescope (EGRET) gathered data on very high-energy gamma rays. Its observations led to the discovery of blazars.

## After whom is the **Compton Gamma Ray Observatory named**?

The CGRO was named for the Nobel Prize-winning American physicist Arthur Holly Compton (1892–1962). He pioneered the study of X-ray reflections in crystals and the scattering of X rays by matter. He also discovered the effect known today as Compton scattering, where X-ray photons transfer some of their energy away when they interact with electrons. (The inverse effect, where subatomic particles add energy to X rays and make the radiation more powerful, is an important way that astronomical objects like quasars produce high-energy X rays and gamma rays.)

## What is the **INTEGRAL satellite**?

The International Gamma Ray Astrophysics Laboratory (INTEGRAL) mission is a gamma ray space telescope operated by the European Space Agency. It was launched from the Baikonur space center in Kazakhstan on a Russian Proton launcher on October 17, 2002. As a successor to COS-B and CGRO, it has several

## What was the final fate of the Compton Gamma Ray Observatory?

The Compton Gamma Ray Observatory, with a mass of nearly 17 tons, was the most massive space telescope ever launched. This meant that, when its orbit degraded and it re-entered the atmosphere, large pieces of metal were likely to survive and crash-land onto Earth's surface. When the CGRO's orbital navigation and control systems began to degrade, NASA officials decided it was too risky to lose complete control of where the pieces might land. On June 4, 2000, CGRO was steered carefully into the atmosphere and successfully de-orbited over the South Pacific, where the pieces fell harmlessly into the ocean.

scientific instruments that make it capable of taking both images and spectra in gamma rays. It also has detectors that can take observations in X rays and visible light at the same time it is taking gamma ray data. This is very useful when studying energetic gamma ray-producing events like stellar explosions or quasar outbursts. Among INTEGRAL's scientific achievements is the creation of a complete map of the sky in low-energy gamma-ray radiation.

## What is the **Swift mission**?

The Swift mission is a mid-sized explorer mission operated by NASA in partnership with Britain and Italy. Swift was launched on November 20, 2004, from Cape Canaveral in Florida. It is a combined gamma ray, X ray, and ultraviolet/visible mission specially designed to study gamma-ray bursts, determining their origins and seeing if they can be used as probes of the early universe. Swift has onboard the Burst Alert Telescope (BAT), a gamma-ray telescope that detects gamma-ray bursts; the X-Ray Telescope (XRT), which narrows down the burst's location by its X-ray emission; and the Ultraviolet-Optical Telescope (UVOT), which takes a detailed image of the location of the burst and whatever leftover afterglow light has been produced by the burst. Swift was designed to take all three kinds of data automatically—within seconds of detecting a gamma-ray burst—as well as send detailed information about the burst immediately to astronomers on the ground so the bursts can be studied instantly. While it is not studying a gamma-ray burst, the Swift telescopes can be used for other scientific investigations, such as a sensitive high-energy X-ray survey of the universe.

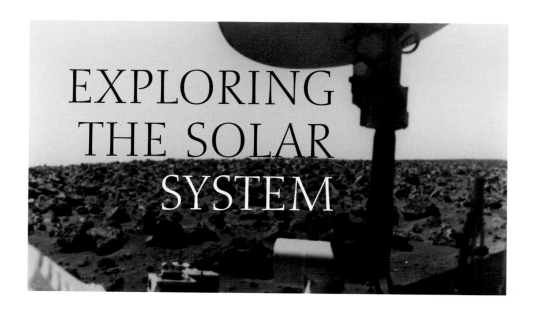

# EXPLORING THE SOLAR SYSTEM

## EXPLORATION BASICS

### What is the **purpose** of sending **spacecraft** to **distant objects**?

Although telescopes on Earth and in Earth's orbit are wonderful instruments for scientific discovery, there is a huge amount of information that cannot be collected from Earth's vicinity. We cannot, for example, use active methods (like hammering, drilling, or even just touching) to get detailed knowledge about a planet's surface or a moon rock. Many surface or atmospheric details are too small for even the largest terrestrial or Earth-orbiting telescopes to resolve. Sometimes, too, the view to a distant object is obscured, so detailed observation is only possible close up.

Spacecraft sent to distant objects also serve another vitally important role: they let humans test and develop advanced technologies that benefit all of humanity. In so doing, they further the human urge to learn, explore, and achieve new things.

### What is a **flyby**?

A flyby is, as its name suggests, a maneuver in which a spacecraft flies past an object in space. Before, during, and after the closest point of approach, the scientific instruments of the spacecraft train on the target object, gathering as much data as possible until it is too far away to gather more.

### What is a **gravitational slingshot**?

A gravitational slingshot, also known as a "gravity assist" or "swing-by," is a special orbital maneuver—a complex, carefully planned flyby—that allows a spacecraft to use the gravitational pull of a solar system object to change the spacecraft's speed and direction. These maneuvers allow mission designers to save fuel and weight, which are two of the most crucial limitations to successful spacecraft operations and mission lifetimes.

## What is an **orbital insertion**?

An orbital insertion is the process of maneuvering a spacecraft into a stable orbit around a solar system object. This kind of maneuver is perhaps the trickiest and one of the most difficult of any space exploration mission because it requires precise timing and tremendous amounts of fuel compared to just about any other maneuver. A little miscalculation can cause the spacecraft to fly off into deep space or crash into the body it is trying to orbit! If an orbital insertion is successful, however, it leads to months or years of detailed scientific data-gathering.

# EXPLORING THE SUN

## Have any **space probes** ever been sent to the **Sun**?

Since the Sun is so bright from our vantage point here on Earth, most of the spacecraft that have been used to study the Sun have been launched only into Earth's orbit. Some of the best-known examples of such spacecraft include the Solar Maximum Mission (SMM), the Solar and Heliospheric Observatory (SOHO), and the Transition Region and Coronal Experiment (TRACE). Some spacecraft, however, have indeed been launched into special orbits around the Sun to study the Sun in ways not possible from Earth. These include the Helios and Ulysses space probes.

## What were the **Helios space probes**?

*Helios 1* was launched on December 10, 1974, and *Helios 2* on January 15, 1976, as a scientific collaboration between the United States and West Germany. These two spacecraft were launched into highly elliptical orbits, where their aphelion (great-

Technicians at Cape Canaveral check out the *Ulysses* before launch. (*NASA*)

est distance from the Sun) was about 1 AU (93 million miles), but their perihelion (closest distance to the Sun) was only 0.3 AU (28 million miles), which is less than the distance to Mercury from the Sun.

Each Helios probe had a number of scientific instruments aboard designed to study the space environment between Earth and the Sun. Among other things, they studied particle emissions from the Sun, the strength of the Sun's magnetic field, zodiacal light, micrometeoroids, and cosmic rays. Their scientific mission ended in the mid 1980s, but the spacecraft are still orbiting the Sun to this day.

### What **interesting record** do the **Helios probes** hold?

The highly elliptical orbits of the Helios spacecraft caused them to change speeds dramatically as they orbited the Sun. At their aphelion distances, they traveled at about 45,000 miles (73,000 kilometers) per hour, while at perihelion their speeds were a remarkable 150,000 miles (240,000 kilometers) per hour. Thus, these two probes hold the record for being the fastest objects ever built in human history. (For the record, *Helios 2* was slightly faster than *Helios 1*.)

### What is the *Ulysses* spacecraft?

The *Ulysses* spacecraft was launched aboard the space shuttle *Discovery* on October 6, 1990, as a joint mission between the European Space Agency and NASA. It was launched at an angle out of the ecliptic plane, toward the planet Jupiter. On February 8, 1992, it then used Jupiter's gravity to slingshot completely out of the ecliptic

Since Mercury is so close to the Sun, any spacecraft that will orbit the planet must be able to withstand brutally high temperatures (often exceeding 800 degrees Fahrenheit [427 degrees Celsius]), light flux (11 times that of Earth), and solar wind from the Sun. Thus, *Messenger* has a ceramic cloth sunshield that will keep the spacecraft cool while it orbits Mercury. *Messenger* is made primarily of graphite epoxy material to make it light, yet strong; and it has a full suite of scientific instruments—including a dual imaging camera system, magnetometer, laser altimeter, and three spectrometers that will study gamma rays and neutrons, infrared and ultraviolet light, and energetic particles and plasma. Fortunately, the closeness of the Sun also means that solar panels can be used to power the systems aboard *Messenger*.

plane and into polar orbit around the Sun. Since then, *Ulysses* has been studying the Sun and the solar system from vantage points that no other solar system object has ever achieved. Aside from gathering unique data about the Sun's polar regions and about solar activity above and below the Sun's poles, *Ulysses* has also been used to study comets like Hale-Bopp and Hyakutake, and it measured the surprising fact that the solar wind from one pole of the Sun is about 100,000 degrees hotter than that from the other pole!

# EXPLORING MERCURY AND VENUS

### What was the **first space probe** sent to **Mercury**?

Little was known about Mercury until the space probe *Mariner 10* photographed the planet in 1975. *Mariner 10* first approached the planet Venus in February 1974, and then used that planet's gravitational field to slingshot itself in the direction of Mercury. The journey from Venus to Mercury took seven weeks. On its first flyby of Mercury, *Mariner 10* came within 470 miles (750 kilometers) of the planet, and photographed about 40 percent of its surface. The probe then continued into orbit around the Sun, flying past Mercury twice more in the next year before running out of fuel.

### What is the **most recent spacecraft** to explore **Mercury**?

For three decades after the successful *Mariner 10* mission, no spacecraft were sent to Mercury. Then on August 3, 2004, NASA launched the *Messenger* (MErcury Surface, Space ENvironment, GEochemistry, and Ranging) mission from Cape Canaveral, Florida, on a Delta II rocket. After three years of travel that included a flyby of Earth and two flybys of Venus, *Messenger* approached Mercury on January 14, 2008.

> ## What were the Venera spacecraft like?
>
> The original *Venera* spacecraft weighed 1,400 pounds (630 kilograms). Its cylindrical body featured a domed top and solar-paneled sides. Attached to one side of the body was an umbrella-shaped radio antenna. The vessels used in successive Venera missions were larger and more complex. *Venera 4* and later models consisted of a carrier vessel and a lander; the last two Venera craft each weighed 8,800 pounds (4,000 kilograms).

It will take three more years and two more flybys of Mercury before *Messenger* finally settles into an orbit around Mercury for a detailed, year-long exploration.

## What has the *Messenger* achieved?

*Messenger* has already made significant new discoveries. During its first flyby of Mercury in January 2008, it took pictures of a side of the planet that has never been seen by human eyes. Scientists had thought, based on *Mariner 10* data gathered in the 1970s, that Mercury was very similar to the Moon, but *Messenger* has shown that Mercury has a unique and active geological history. The surface has long fault lines, a remarkable spider-like formation in the center of its Caloris basin, and also has significant pressure within its magnetosphere.

## What were the **first spacecraft** sent to study **Venus**?

The first spacecraft sent to explore Venus were the probes of the Venera program. Venera, the Russian word for "Venus," was an intensive effort by the former Soviet Union to explore the planet between 1961 and 1983. In all, sixteen Venera spacecraft were launched. During that time, the exploration of Venus was almost exclusively the domain of the Soviets.

## What is the **history** of the **Venera program**?

The Venera program had a rocky start. The first three Venera missions, launched between 1961 and 1965, were unsuccessful. Radio contact with the first two probes was lost long before the craft reached Venus, and *Venera 3* crash-landed on the surface of the planet. After that, though, the program began to succeed in gaining scientific results. *Venera 4* reached Venus on October 18, 1967, and released its lander capsule successfully; it broadcast scientific data on the Venusian atmosphere for 94 minutes before it was crushed by the intense atmospheric pressure. Launched on August 17, 1970, *Venera 7* successfully landed on the surface of Venus on December 15, the first successful soft landing of a spacecraft on another planet. This capsule was equipped with a cooling device that helped it survive for 23 minutes after landing. *Venera 8* survived for 50 minutes. *Venera 11, 12, 13,* and *14* all had successful landings, too. The various landers measured, among many other things, the amount of sunlight that reached the surface of Venus, the chemical composition of the atmosphere and surface rocks, and the presence of lightning in the planet's atmosphere.

*Venera 15* and *16* arrived at Venus in October 1983. Rather than drop probes to the surface, they remained in orbit and constructed detailed maps of the surface of Venus using Doppler radar systems. Over the next year, they mapped a large part of the northern hemisphere, including areas that were probably active with volcanoes in the distant past.

## What was the **Vega program**?

The Vega program was a pair of space probes launched six days apart by the Soviet Union in December 1984. They had two destinations: Venus and Comet Halley. Each Vega spacecraft was about 36 feet (11 meters) long and consisted of a cylindrical mid-section with a landing capsule at one end, a communications antenna and solar panels protruding from its central portion, and an experiment platform at the other end. The platform held scientific instruments provided by many nations, including the Soviet Union, France, Germany, and the United States. Although this is a common practice today, at the time it was a pioneering example of international cooperation in space exploration.

## What did the **Vega probes achieve**?

*Vega 1* flew by Venus on June 11, 1985, and dropped a science capsule and a high-altitude balloon-borne payload to the Venusian surface. The capsule landed safely and relayed pictures and other scientific data for two hours. At the same time, the helium-filled balloon carrying scientific instruments hovered in the atmosphere of Venus for two days at an altitude of about 31 miles (50 kilometers). During that time, the balloon was blown more than 6,200 miles (10,000 kilometers) from its original position. The instruments gathered valuable scientific data about the temperture, pressure, and wind speeds of the Venusian atmosphere. The entire capsule-and-balloon scientific sequence was repeated a few days later by *Vega 2*.

After releasing their scientific payloads at Venus, the Vega spacecraft then used Venus as a gravitational slingshot, propelling them on an intercept course with Comet Halley. On March 6, 1986, *Vega 1* came within 5,600 miles (9,000 kilometers) of the comet's nucleus; *Vega 2* had its closest approach three days later. The two probes collected substantial scientific data on the comet; some of the data were used by the European Space Agency to reposition the *Giotto* probe to the comet. After passing Comet Halley, the Vega spacecraft remained in orbit around the Sun until they were shut down in early 1987.

## What **probes** did the **United States** send to **Venus** in the **1960s and 1970s**?

The first U.S. spacecraft to reach Venus successfully was *Mariner 2,* which flew by Venus in 1962. Another flyby was achieved by *Mariner 10* in 1974; as the spacecraft headed toward Mercury, it took many close-up pictures.

In 1978, the United States launched two more spacecraft to explore Venus. The first, called *Pioneer-Venus Orbiter* (PVO), was launched on May 20, 1978. It studied the planet's atmosphere and mapped about 90 percent of the Venusian surface. It also made observations of several comets that passed near Venus, and provided information on mysterious gamma-ray bursts. PVO ran out of fuel in October 1992, and it descended into the Venusian atmosphere and burned up. The *Pioneer-Venus Multiprobe* (PVM) was launched on August 8, 1978, and distributed four probes around the planet, which then traveled down through the atmosphere and onto the surface of Venus. They measured atmospheric temperature, pressure, density, and chemical composition at various altitudes. One of the four probes survived after impact; it transmitted data from the surface for 67 minutes.

The *Magellan* is attached to a booster rocket before its launch in 1989. *(NASA)*

## What was the *Magellan* mission to **Venus**?

The *Magellan* spacecraft, named after the sixteenth-century Portuguese explorer, was launched by NASA on May 4, 1989; it was the first scientific spacecraft to be launched from a space shuttle, the *Atlantis*. The spacecraft reached the planet on August 10, 1990. *Magellan* was equipped with a sophisticated Doppler radar mapping system, which astronomers used, along with altimetry and radiometry data, to measure and map the planet with unprecedented accuracy. *Magellan* ultimately made a three-dimensional map of 98 percent of the surface of Venus, and was able to measure features with a precision of 100 meters (330 feet).

After concluding its radar mapping, *Magellan* transmitted a constant radio signal. By measuring changes in the frequency of the signal as *Magellan* orbited, astronomers were able to use the spacecraft to make global maps of Venus's gravity field. After four successful years of the scientific study of Venus, the *Magellan* mission ended on October 11, 1994. Flight controllers flew the spacecraft into the atmosphere and onto the planet's surface, the first time an operating planetary spacecraft had ever been intentionally crashed.

> ## What unique maneuver
> ## did the *Magellan* accomplish while orbiting Venus?
>
> Flight controllers also used *Magellan* to test a new maneuvering technique called aerobraking, which uses a planet's atmosphere to slow or steer a spacecraft. This technique became very important in the development of future planetary lander spacecraft.

## What is the *Venus Express*?

The *Venus Express* mission is a project designed and operated by the European Space Agency. It was launched from the Baikonur Cosmodrome in Kazakhstan on November 9, 2005, on a Soyuz-Fregat rocket. It arrived at Venus on April 11, 2006, eventually settling into a 24-hour elliptical, quasi-polar orbit.

## What has the *Venus Express* accomplished so far?

With its suite of scientific instruments, which includes cameras, spectrometers, and a magnetometer, *Venus Express* has studied the atmosphere, electromagnetic characteristics, and surface of Venus in great detail. It has made important strides toward understanding the origin of the runaway greenhouse effect on Venus; studied the Venusian surface and atmosphere using infrared light; and may have confirmed the occurrence of lightning on the planet.

# EXPLORING MARS

## What was the **Soviet Union's Mars program**?

The Soviet Union was the first nation to send spacecraft to Mars. After a number of unsuccessful tries, they launched the *Mars 1* spacecraft in late 1962, but lost radio contact with it after a few months. In 1971 the Soviets successfully put *Mars 2* and *Mars 3* into orbit around Mars. Both of these craft carried landing vehicles that successfully dropped onto the Martian surface. Unfortunately, in each case radio contact was lost after only a few seconds. In 1973 the Soviets sent out four more spacecraft toward Mars, one of which successfully transmitted back data about the Red Planet.

## What was the Soviet Union's **Phobos** program?

In 1988, the Soviet Union renewed its interest in exploring Mars. They sent two identical spacecraft, *Phobos 1* and *Phobos 2*, both headed for the larger Martian moon Phobos. Unfortunately, contact with both probes was lost before either reached its destination.

An image of Mars's Utopia Planitia taken by *Viking 2*. (*NASA*)

## What were the first American **Mariner** missions to **Mars**?

The first U.S. probe to Mars, *Mariner 4*, flew past Mars on July 14, 1965. It sent back 22 pictures of the Red Planet, and gave us our first glimpse of its cratered surface. It also detected the thin Martian atmosphere, made mostly of carbon dioxide and less that one percent the density of the atmosphere on Earth. In 1969, *Mariner 6* and *Mariner 7* flew by Mars, and together they produced 201 new images of Mars, as well as more detailed measurements of the structure and composition of the Martian surface, polar caps, and atmosphere.

## What was the **first spacecraft** to **orbit Mars**?

In 1971 *Mariner 9* became the first spacecraft to enter orbit around Mars. During its year in orbit, *Mariner 9* sent back pictures of an intense Martian dust storm, as well as images of 90 percent of the planet's surface and of Phobos and Deimos, the two Martian moons.

## What were the **first spacecraft** to **land safely** on **Mars**?

In 1976, two spacecraft sent by the United States arrived at Mars. *Viking 1* arrived on June 19, and *Viking 2* arrived on August 7. Each spacecraft had an orbiter and a lander. The *Viking 1* lander touched down on Mars in the Chryse Planitia on July

271

20, followed by the *Viking 2* lander in the Utopia Planitia (Utopian Plane) on September 3. The orbiters sent back detailed pictures, radiation measurements, and weather reports of the entire Martian surface, and the landers obtained the first images ever taken from the surface of another planet.

### How were the **Viking spacecraft configured**?

Each Viking craft had two parts, an orbiter and a lander. Each orbiter was an eight-sided structure about eight feet (2.4 meters) wide. Most of the spacecraft's control systems were contained in this body. The rocket engine and fuel tanks were attached to the rear face of the structure. Solar panels extended from another face. These panels were extended once in space to a cross-shaped structure about 32 feet (10 meters) across. The orbiter also contained a movable platform on which scientific equipment was mounted, including two television cameras and instruments for measuring the temperature and water content of the Martian surface.

Together, the lander and orbiter stood 16 feet (5 meters) tall. The central portion of each lander was a six-sided compartment with alternating longer and shorter sides. Attached to each short side was a landing leg with a circular footpad. A remote control arm for the collection of soil samples, which resembled an extended, pointy fourth leg, protruded from one of the lander's long sides. Soil samples were transferred to the biological analyzer, which was perched on top of the body, for testing and analysis. Other instruments affixed to the top of the lander included two cylindrical television cameras, a seismometer for measuring Mars-quakes, atmospheric testing devices, and a radio antenna dish. Beneath the lander were rockets that slowed the lander's descent, and the propellant for the rockets were stored in tanks on opposite sides of the lander.

### What was the **first mobile mission** ever sent to **another planet**?

*Mars Pathfinder,* which was launched on December 4, 1996, landed on the planet Mars on July 4, 1997. It carried with it the *Sojourner* mobile unit, or "Mars buggy," which rolled out of the *Pathfinder* craft soon after the landing. *Sojourner* soon made history as the first robotic vehicle to move under its own power across the surface of another planet.

### What did *Pathfinder* see on **Mars**?

The *Pathfinder* spacecraft was equipped with a stereoscopic camera on a 360-degree rotating mount, with two lenses mounted several inches apart that allowed scientists to take both detailed zoomed-in images as well as three-dimensional panoramic views of Mars. *Pathfinder* saw, moreover, that the Red Planet's sky looks kind of pinkish-yellowish-red from the surface, thanks to varying amounts of dust particles that are always present in the atmosphere. In all, *Pathfinder* took more than 16,500 digital photos.

## What is the **Carl Sagan Memorial Station**?

The *Pathfinder* lander was eventually renamed the Carl Sagan Memorial Station, in honor of the American astronomer who had helped popularize astronomy and astrophysics in the late-twentieth century and inspired an entire generation of space scientists.

The *Pathfinder* is prepared at the Launch Complex 17B on Cape Canaveral Air Station for its Mars mission. (*NASA*)

## How did the *Sojourner* rover work?

The *Sojourner* rover, or "Mars buggy," was a mere one foot high, two feet long, and a foot and a half wide. It rolled out of the *Pathfinder* spacecraft and was capable of traveling several feet per day on its six wheels. It ran on batteries charged by its solar panels, which converted energy from the Sun into electrical power, and it was controlled remotely by scientists on Earth.

## How **long** did *Mars Pathfinder* and the *Sojourner* rover **last**?

The *Pathfinder* probe and *Sojourner* rover each operated for about three months. This far exceeded the mission's original specifications; *Pathfinder* was expected to work for 30 days, and *Sojourner* for just seven days. The mission netted more than 17,000 images—550 of them from the *Sojourner*'s mobile camera—and 2,300 megabytes of data.

## What was **pioneering** about the *Pathfinder* mission **strategy**?

*Mars Pathfinder* was an overwhelming success as the first of a series of space probes designed, as then-NASA administrator Dan Goldin described, to be "faster, better, and cheaper." At a cost of approximately $200 million, it was about one-twentieth the cost of the Viking spacecraft that preceded *Pathfinder* to Mars two decades before. By using creative strategies like the "bounce landing," and by taking calculated risks with the technology, the Pathfinder program showed that it was possible to get a high scientific return for relatively small cost. It began a trend in space exploration that moved away from the previous model of single higher-cost, high-complexity spacecraft, and toward a model of multiple, lower-cost missions to achieve the same scientific goals.

273

A 360-degree view of Mars taken by the *Pathfinder*. (*NASA*)

### What **equipment** did the *Sojourner* rover carry?

*Sojourner* had several scientific instruments onboard, such as an alpha-proton X-ray spectrometer, which was used to analyze the chemical composition of soil and rocks that it encountered. Several of the rocks it encountered had distinctive shapes, which led the Pathfinder scientists to give them colorful names like "Yogi" and "Barnacle Bill."

### What was the *Mars Global Surveyor*?

The *Mars Global Surveyor* (MGS) was an orbiter sent to explore Mars and send back information about how the Martian climate and landscape have been changing over time. It was launched by NASA from Cape Canaveral on November 7, 1996, on a Delta-7925 rocket, and arrived at Mars on September 11, 1997. It had actually been built using many spare parts from the failed *Mars Observer* mission. There was initial concern that the fate of MGS was doomed as well, when one of its two solar panels did not deploy properly after launch.

Using the pioneering method of "aerobraking"—adjusting and slowing the spacecraft's orbit by flying it through the top of the Martian atmosphere—MGS was gently placed into a nearly circular orbit over the course of more than a year. MGS began mapping the surface of the Red Planet in March 1999. By carefully managing its fuel and electrical energy supply, scientists were able to extend the life of MGS more than five years beyond its primary mission. The *Mars Global Surveyor* was lost in November 2006, after taking more pictures than all other Mars missions.

# FAILED MARS MISSIONS

### What was the *Mars Observer* mission?

Since the Viking mission, a number of scientific missions to Mars have ended in failure. One particularly high-profile failure was NASA's *Mars Observer* mission. Launched on September 25, 1992, it was a $1 billion mission that had a tremendous amount of scientific capability. Unfortunately, on August 21, 1993, three days before it was to enter orbit around Mars, flight engineers lost contact with *Mars Observer*. It was never heard from again. Investigators suspect that ruptured tubing in the spacecraft's propulsion system caused it to spin out of control.

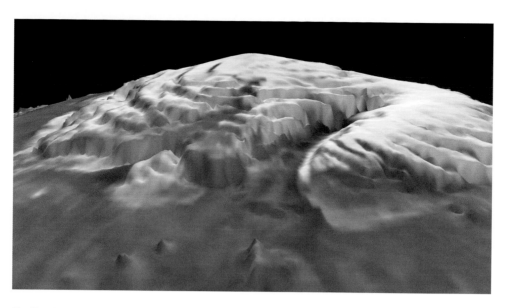

The *Mars Global Surveyor* took this first-ever three-dimensional image of the north pole of Mars. (*NASA*)

### What was the *Mars 96* mission?

After the collapse of the Soviet Union, the Russian space program had designed *Mars 96* as the centerpiece program of the new Russian Space Agency. It consisted of an orbiter, two small space stations for landing on the Martian surface, and two penetrating probes to bore into the surface and examine the underground environment of the planet. The spacecraft carried numerous scientific instruments for studying the Martian surface, atmosphere, and magnetic field. Launched on November 16, 1996, *Mars 96* unfortunately failed to reach orbit. As the spacecraft flew over the Atlantic Ocean, the launcher's fourth rocket stage failed to fire. The spacecraft crashed into the south Pacific, near South America. The Russian Space Agency has yet to launch another mission to Mars.

### What was the *Nozomi* mission?

The Japanese Space Agency launched the *Nozomi* ("Hope") mission to Mars on July 4, 1998. The planned 15-month long trip to Mars immediately ran into trouble when a thruster failed early in the mission. For five long years, scientists and engineers worked to keep the spacecraft on its way toward the Red Planet. By 2003 it looked like it might make it after all, but all hope was lost when, on December 9, 2003, flight controllers were unable to orient the spacecraft properly for orbital insertion. *Nozomi* flew by Mars at a distance of about 630 miles (1,000 kilometers), and was propelled into orbit around the Sun.

### What were the *Mars Climate Orbiter* and *Mars Polar Lander*?

On December 11, 1998, NASA launched the first of a pair of Mars explorer probes, the *Mars Climate Orbiter* (MCO). On January 3, 1999, the *Mars Polar Lander* (MPL) was successfully launched, as well. The plan was for MCO to enter orbit around Mars in October 1999 and begin transmitting signals back to Earth; MPL would then land in December, examine the surface of Mars for signs of liquid water, and other substances, and relay the information back to Earth via the MCO.

But on September 23, 1999, as MCO fired its main thrusters to insert into orbit around Mars, flight controllers lost contact with the spacecraft. After a frantic investigation, it was discovered that the wrong amount of rocket thrust had been applied (the navigation software had used the wrong mathematical units of force in its calculations). This simple, careless human error had caused MCO to crash onto the Martian surface. Scientists quickly sent the corrected information to the *Mars Polar Lander,* to be sure it would not suffer the same fate.

On December 3, 1999, as MPL came down for a soft landing, flight controllers lost contact with it less than 12 minutes before it was to touch down near the Martian south pole. Later investigations suggested that the engines had erroneously shut down while MPL was still more than 100 feet in the air, causing a devastating crash landing.

# MARS MISSIONS IN THE TWENTY-FIRST CENTURY

### What is *2001 Mars Odyssey*?

The *2001 Mars Odyssey* mission—so named in part to honor the classic science fiction novel *2001: A Space Odyssey* by Arthur C. Clarke—was launched by NASA on April 7, 2001, on a Delta II rocket, and successfully inserted into orbit on October 24, 2001. Equipped with three main scientific instruments—THEMIS (Thermal Emission Imaging System), GRS (Gamma Ray Spectrometer), and MAREE (Mars Radiation Environment Experiment)—*2001 Mars Odyssey* successfully completed its primary scientific mission between February 2002 and August 2004. It began its extended mission on August 24, 2004.

### What has *2001 Mars Odyssey* achieved?

Aside from studying the properties of Mars—primarily its climate, geological history, and its potential to support life as we know it—*Mars Odyssey* has been very

> ## What has been the most
> ## significant scientific discovery made with Mars orbiters?
>
> **B**y far the most significant scientific discovery made with Mars orbiters has been conclusive evidence that Mars once had liquid water on its surface, and still has liquid water underground today. Using methods to search for various substances on and in the Martian crust—the same way that surveyors use satellites on Earth—astronomers have confirmed the existence of Martian rocks that could only have formed in the presence of water; chemical evidence for the past presence of water; the remains of recent geyser-like activity where water has spurted out from cracks and fissures in canyon walls; and even a vast underground sea of ice that is larger than Pennsylvania, Ohio, Indiana, Kentucky, and Illinois put together!

important in the search for appropriate landing sites for future Mars exploration missions. It also has a powerful communications array, and it continues to serve as the primary communications relay station between scientists and flight controllers on Earth and the Mars Exploration Rovers *Spirit* and *Opportunity*.

## What is the *Mars Express* orbiter?

The *Mars Express* was a mission sent by the European Space Agency to Mars. It was built by a consortium of 15 nations led by France, and consisted of an orbiter and a lander called *Beagle 2*. *Mars Express* was launched on June 2, 2003, from the Baikonur launch site in Kazakhstan onboard a Russian Soyuz/Fregat rocket. It arrived into orbit around Mars on Christmas Day. Six days before its arrival, it released the *Beagle 2* toward the Martian surface; unfortunately, the lander was lost, and has yet to be found.

The *Mars Express* orbiter, happily, has been completely successful. It has lasted far longer than its originally planned mission lifetime of two years, and continues to take detailed images and other data of Mars, and to serve as a commications relay for data from other Martian missions.

## What is the *Mars Reconnaissance Orbiter*?

The *Mars Reconnaissance Orbiter* (MRO) was launched on August 12, 2005, from Cape Canaveral, Florida, on an Atlas V-401 rocket. It arrived without a hitch at Mars on March 10, 2006, and spent the next six months aerobraking from its initial, highly elliptical orbit into its final, nearly-circular orbit. MRO has taken the most detailed pictures from Martian orbit yet of geological features and conditions on the surface of Mars. It has also been successfully used as a communications relay for other scientific missions to Mars—an important task for future years. As of November 2007, MRO had already returned more than 26,000 gigabytes of data—more than all other Mars missions put together.

## What is the **Mars Exploration Rover** (MER) program?

The Mars Exploration Rover (MER) program was built on the tremendous success of the *Mars Pathfinder* mission to explore the surface of Mars using mobile, remotely-controlled robotic rovers. The two spacecraft of the MER program, the rovers *Spirit* and *Opportunity,* succeeded beyond all expectations, giving us a remarkable and inspiring view of the geology of the Red Planet.

## How and when did *Spirit* and *Opportunity* land on Mars?

*Mars Exploration Rover A,* known as *Spirit,* launched from Cape Canaveral, Florida, on June 10, 2003, and arrived at Gusev Crater on January 3, 2004. *Mars Exploration Rover B,* known as *Opportunity,* launched on July 7, 2003, and arrived at the Meridiani Planum on January 25, 2004, on the opposite side of Mars from *Spirit.* Like the *Mars Pathfinder* mission before it, *Spirit* and *Opportunity* landed on Mars by slowing down from 12,000 miles per hour to 12 miles per hour by parachute and rocket, then tumbling and bouncing across the Martian surface on huge, 18-foot-high cushioned air bags until they came to rest. Both landings were completely successful.

## How were the Mars probes *Spirit* and *Opportunity* configured?

The Mars probes *Spirit* and *Opportunity* were designed to be robotic geologists. The size of small golf carts, the twin probes are about as heavy as, and move around with the speed of, a Galapagos tortoise—about 130 feet (40 meters) a day. They were equipped with many of the instruments that a geologist on Earth might carry on a scientific expedition. *Spirit* and *Opportunity* were remotely controlled by scientists on Earth, but were also programmed with a small amount of autonomy to adapt to immediate conditions that they encountered on Mars.

## What other **instruments** are on the *Spirit* and *Opportunity* probes?

The primary instruments on *Spirit* and *Opportunity* include a stereoscopic Panoramic Camera for studying the local and distant terrain; a miniature thermal emission spectrometer (Mini-TES) for identifying rocks and soils and for looking skyward to measure the Martian atmosphere; a Moessbauer Spectrometer for close-up mineralogical studies of iron-bearing rocks and soil; an alpha-particle X-ray spectrometer to measure the chemical composition of rocks and soil; magnets for collecting dust particles; and a Microscopic Imager for obtaining close-up, super-high resolution pictures of Martian rocks and soil.

## What were some of the **highlights** of the travels of the *Spirit* rover?

*Spirit* landed in a rocky, flat area of the large Gusev Crater, which was thought to be a possible lakebed that had dried up millions, or even billions of years ago. Its landing site was named Columbia Memorial Station, in honor of the lost space shuttle and its crew. *Spirit* studied a number of nearby rocks—nicknamed, among others, "Adirondack," "Mimi," and "Humphrey"—and uncovered strong evidence that the geology of the region had been shaped long ago by the presence of liquid water.

---

**What is a RAT doing on Mars?**

The RAT is the Rock Abrasion Tool—a nifty piece of scientific equipment on the Mars rovers *Spirit* and *Opportunity*. About the size of a human hand, the RAT can delicately grind away the paper-thin outer layers of Martian rocks that the rovers encounter. This allows scientists to study the inner portions of those rocks that have not been altered by weather or radiation.

---

*Spirit* moved to the Bonneville Crater, 400 yards away from Columbia Station, and then spent the next two years moving toward and studying the Columbia Hills several miles away.

Overall, *Spirit* has traveled about five miles from Columbia Station. About three years into the mission, it established a new, temporary base on a rocky plateau nicknamed "Home Plate." By staying on a north-facing slope, it will hopefully be able to continue getting enough sunlight to its solar panels to survive the Martian winter.

## What were some of the **highlights** of the travels of the *Opportunity* rover?

*Opportunity* landed in the broad expanse of Mars' Meridiani Planum, right in the middle of a small, 60-foot wide crater. The crater was named Eagle Crater in honor of the *Apollo 11* mission to the Moon. After studying some of the geological features within the Eagle Crater—which were given nicknames like "Stone Mountain," "El Capitan," and "Opportunity Ledge"—*Opportunity* dug a shallow trench into the Martian soil by spinning one of its wheels to look at the soil underneath the surface.

*Opportunity* then made its way to Endurance Crater, exploring it for six months before backing out again. As it traveled, it encountered Heat Shield Rock, the first meteorite ever identified on a celestial object other than Earth. In April 2005, more than two years after the landing, the rover accidentally got stuck in a sand dune, which mission scientists dubbed "Purgatory Dune." It took nearly two months of delicate planning and maneuvering to get *Opportunity* out of Purgatory, but it finally escaped on June 4, 2005. It then moved to study Victoria Crater, four miles from the landing site. *Opportunity* has traveled more than seven miles since its landing. It holds the record for the longest rover travel in a single day: 582 feet (177.5 meters).

# EXPLORING THE OUTER PLANETS

## What was the **Pioneer program**?

The Pioneer program of space probes was begun in 1958 by the U.S. Department of Defense and the newly formed NASA. They were designed to travel beyond Earth's orbit and gather scientific data about the objects in the solar system.

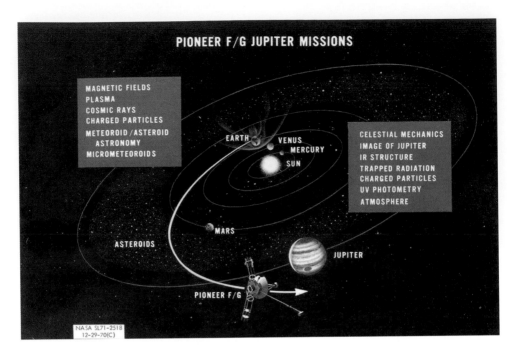

Before computers were used, artists were hired by NASA to illustrate planned missions, such as this 1970 plan for the Pioneer 10, which lists some of the planned measurements and photos the probe would take of the solar system. (*NASA*)

## What were the **first Pioneer probes**?

The first three Pioneer probes were drum-shaped spacecraft weighing 84 pounds (38 kilograms) each. They were intended to go into orbit around the Moon. Unfortunately, they failed to leave Earth's gravity. *Pioneer 4* was a much smaller payload at just 13 pounds (six kilograms), and was designed to fly by the Moon rather than orbit it. *Pioneer 4* was successfully launched out of Earth's gravitational pull, and passed within 37,000 miles (60,000 kilometers) of the Moon, but it was too far away to gather any scientific data. Four more moon probes were unsuccessful, too.

## What did *Pioneer 5* through *9* accomplish?

*Pioneer 5* was the first of five Pioneer probes launched into an orbit around the Sun. Generally, a solar orbit is far easier to achieve than a lunar orbit because it can be a much less precise task: any object launched beyond Earth's gravitational influence will naturally tend to go into solar orbit, unless carefully aimed to do otherwise. *Pioneer 5* was launched on March 11, 1960, and was a sphere about 25 inches (64 centimeters) across and weighed 95 pounds (43 kilograms). It was the first satellite to maintain communications with Earth at the then-impressive distance of 23 million miles (37 million kilometers). *Pioneers 6* through *9* were successfully launched into solar orbit between 1965 and 1968. Each weighed about 140 pounds (64 kilograms), was covered with solar cells, and carried instruments to measure cosmic rays, magnetic fields, and the solar wind. Overall, these five Pioneer spacecraft lasted in solar orbit for decades.

> ## What is the oldest probe in the solar system?
>
> The *Pioneer 6* is still considered "extant," and is the oldest operating probe in the history of space exploration.

## What were *Pioneer 10* and *Pioneer 11* designed to do?

The two best-known Pioneer spacecraft, *Pioneer 10* and *11*, left Earth in 1972 and 1973 respectively. They were designed to gather data about the distant gas giant planets Jupiter and Saturn. Each of the twin spacecraft had a nine-foot (three-meter) diameter radio antenna dish, which were used for communications between the spacecraft and receiving stations on Earth. Scientific instruments, cameras, a radioisotope thermoelectric generator (RTG), and a rocket motor were attached to the back of the dish.

## What **milestones** did *Pioneer 10* accomplish?

*Pioneer 10* was the first spacecraft ever to cross the asteroid belt. Before this, astronomers did not have a clear idea whether the density of tiny asteroids in the belt would be too great for a ship to go through it without being smashed. (It was not—the nearest *Pioneer 10* came to a known asteroid was 5,500,000 miles [8.8 million kilometers].) In 1973 *Pioneer 10* flew by Jupiter and took the first close-up pictures of the largest planet in our solar system. It then kept traveling, crossed the orbits of Neptune and Pluto, and left the major planet region of the solar system in 1983. The last successful contact with *Pioneer 10* was made on January 23, 2003. At its current rate and direction of travel, it will reach the star Aldebaran in the constellation Taurus in about two million years.

## What **milestones** did *Pioneer 11* accomplish?

*Pioneer 11*, like *Pioneer 10*, headed first to Jupiter, taking breathtaking pictures and gathering scientific data on that planet. Then, *Pioneer 11* used Jupiter's gravitational field to slingshot itself toward Saturn. It arrived at Saturn in 1979 and took the first close-up images and scientific data about that planet, its rings, and its moons. In 1990, *Pioneer 11* exited the major planet region of the solar system, and in September 1995, after 22 years of operation, it ran out of power and the routine daily mission operations stopped. The last successful communication with *Pioneer 11* occurred in November 1995.

## What was the **Voyager program**?

The Voyager probes were originally going to be named *Mariner 11* and *Mariner 12*, respectively. They were then moved into a separate program named Mariner Jupiter-Saturn, and renamed *Voyager*.

**R**ight now, *Voyager 1* is more than 105 astronomical units (9.8 billion miles) away from Earth, making it the most distant man-made object in the solar system—in fact, the most distant known object in the solar system, period! Scientific evidence appears to show that the spacecraft has reached the inner edge of the heliosheath—the tear-shaped shell of charged particles that surrounds the solar system as it orbits within the Milky Way. If this is so, *Voyager 1* will probably reach the heliopause—the outer edge of the heliosheath—in less than a decade, making it the first true interstellar space probe.

*Voyager 2* is about 85 astronomical units (7.9 billion miles) away from Earth, heading in a direction roughly perpendicular to *Voyager 1*. Although it is closer to Earth than *Voyager 1,* it is still more than twice as far away as Pluto. In December 2007, measurements taken by *Voyager 2* suggest that the heliosphere is slightly deformed by the Milky Way's interstellar magnetic field—that is, the solar system is dented in its southern side.

The two Voyager probes were timed and designed to take advantage of a special alignment of the planets that would allow a single spacecraft to reach all of the gas giant planets through a series of gravitational slingshots. The original ambitious plan, called the "Grand Tour Program," was scaled back to these two probes due to budget cuts, but nonetheless achieved its primary scientific goal with flybys of Jupiter, Saturn, Uranus, and Neptune.

## What **milestones** did *Voyager 1* accomplish?

*Voyager 1* was launched on September 5, 1977, from Cape Canaveral, Florida, on a Titan 3E Centaur rocket. Even though it was launched a few days after *Voyager 2,* it was sent on a faster trajectory to the outer solar system and thus arrived first. In 1979 *Voyager 1* passed by Jupiter and took pictures of the planet's swirling clouds and Galilean moons. It discovered volcanic activity on Io, and found a previously undiscovered ring around Jupiter. At its closest approach on March 5, it was 217,000 miles (349,000 kilometers) from the center of the planet.

*Voyager 1* successfully used Jupiter as a gravitational slingshot to get to Saturn. It flew by Saturn in November 1980, with the closest approach on November 12 of 77,000 miles (124,000 kilometers). It detected the complex structure of Saturn's rings, and studied the thick atmospheres of Saturn and its moon Titan. When it flew by Titan, the gravitational slingshot it received flung the spacecraft out of the ecliptic plane, sending *Voyager 1* upward and away from the planets.

## What **milestones** did *Voyager 2* accomplish?

*Voyager 2* was launched on August 20, 1977. Like *Voyager 1,* it lifted off from Cape Canaveral, Florida, on a Titan 3E Centaur rocket. It reached its closest approach to

Jupiter on July 9, 1979, at a distance of 350,000 miles (570,000 kilometers). It confirmed and observed volcanic activity on Jupiter's moon Io, as well as crisscrossing lines on the surface of Europa; it also discovered several new rings and three new moons around Jupiter, and studied the Great Red Spot in detail.

Using the Jupiter flyby as a gravitational slingshot, *Voyager 2* made it to Saturn; its closest approach was on August 25, 1981. It probed the upper atmosphere of Saturn with its radar system, and took pictures of Saturn, its rings, and its moons. It then used the Saturn flyby to slingshot it to Uranus, reaching its closest approach of 81,500 kilometers (50,600 miles) on January 24, 1986. It discovered 10 previously unknown moons of Uranus, and studied the Uranian moons, atmosphere, magnetic field, and thin ring system.

Finally, using the flyby of Uranus as a gravitational slingshot, *Voyager 2* made its closest approach to Neptune on August 25, 1989, only 3,000 miles (4,800 kilometers) above the planet's north pole. Scientists expected it to find a planet very similar to Uranus. Instead, it found a dynamic, bluish-hued atmosphere covered with some of the strongest winds and storms in the solar system. *Voyager 2* also found four partial ring arcs and six new moons around Neptune; measured the length of Neptune's day and the strength of its magnetic field; and studied Neptune's moon Triton in great detail, revealing a thin atmosphere, clouds, polar caps, and remarkable volcanic geysers of pressurized water.

## What was the *Galileo* mission?

*Galileo* was a major, multi-billion dollar mission to Jupiter and its four largest moons: Io, Europa, Ganymede, and Callisto. Along the way, it also tested space probe flight strategies—in particular, large-scale gravitational slingshots—and even studied Earth as if it were a distant planet. This highly successful mission overcame a great many obstacles to achieve and exceed its scientific expectations; it was therefore called "the little spacecraft that could."

## How was the *Galileo* spacecraft configured?

About the size of a minivan with a flagpole sticking out of its side, *Galileo* weighed two and a half tons fully loaded. It contained a suite of scientific instruments, two communications antennae, thruster rockets, and radioisotope ther-

The *Galileo* probe made an extensive survey of Jupiter and its Galilean moons. (*NASA*)

The *Galileo* spacecraft was originally designed to launch from the space shuttle and be pushed by a powerful booster rocket toward Jupiter. But just a few months before its scheduled launch, the space shuttle *Challenger* exploded in midair, causing a full stop to the shuttle program. Safety concerns dictated that the booster rockets to be used on all future shuttle flights had to be much smaller and less powerful than the one to be used by *Galileo*. Faced with this obstacle, *Galileo* scientists were forced to recalculate the spacecraft's trajectory to Jupiter, using several flybys of Venus and Earth as gravitational slingshots and lengthening the journey by years. Finally, on October 18, 1989, *Galileo* was launched aboard the space shuttle *Atlantis,* and began its six-year journey to Jupiter.

moelectric generators to provide power. *Galileo*'s mini-probe, about the size of a dishwasher, contained six scientific instruments of its own designed to measure the conditions of its surroundings as it descended by parachute into Jupiter's atmosphere.

## What was *Galileo*'s flight path to Jupiter?

*Galileo* needed three major gravitational slingshots to gather enough speed to make it to Jupiter. The VEEGA (Venus-Earth-Earth Gravity Assist) maneuver caused *Galileo* to fly by Venus on February 10, 1990; Earth on December 8, 1990; and Earth again on December 8, 1992. The extra flight time and distance proved to be scientifically fortuitous. *Galileo* was able to pass close by and thus study two asteroids: Gaspra (on October 29, 1991) and Ida (August 28, 1993). On the latter pass, it found the first-ever moon around an asteroid: the smaller asteroid Dactyl, orbiting Ida. Then in 1994, about a year away from its destination, *Galileo*'s cameras were well positioned to observe the collision of the fragments of Comet Shoemaker-Levy 9 into Jupiter.

## How did the *Galileo* spacecraft's mini-probe work?

On December 7, 1995, *Galileo*'s mini-probe dropped from the spacecraft and entered Jupiter's atmosphere at a speed of 106,000 miles (170,000 kilometers) per hour. Within two minutes, it had slowed to less than 110 miles (170 kilometers) per hour. Soon after, the probe deployed a parachute, which slowed its descent further, and it floated downward toward Jupiter's core. As it went down, intense winds blew it nearly 300 miles (500 kilometers) horizontally. In all, the mini-probe lasted for 58 minutes, taking detailed pictures and measurements of the giant planet until its instruments stopped working about 90 miles (150 kilometers) below the top of Jupiter's atmosphere. Eight hours later, the probe vaporized as temperatures reached 3,400 degrees Fahrenheit (1,900 degrees Celsius).

> ### What experiment did the *Galileo* conduct in its 1992 flyby?
>
> **D**uring the December 1992 flyby of Earth, *Galileo* conducted another experiment to see if visible-light lasers could be used to communicate with spacecraft. The Galileo Optical Experiment (GOPEX) was successful; scientists on Earth fired a series of bright laser pulses from ground stations in California and New Mexico, and Galileo took digital pictures of the pulses, successfully detecting about one-third of them up to a distance of nearly four million miles.

## How long did *Galileo* operate?

*Galileo* successfully achieved Jovian orbit on December 7, 1995. Unfortunately, its high gain antenna had failed, so astronomers had to receive data from a much weaker backup antenna. Working creatively, scientists were able to increase the communication speed by nearly tenfold; but even at its best, the transmission rate was still only one percent the speed of a dial-up modem here on Earth.

*Galileo* outlasted even optimistic estimates of its lifetime. After its primary science mission ended two years after orbital insertion, the spacecraft continued its extended mission for more than five years afterward. *Galileo*'s cameras finally succumbed to radiation damage, and were shut off on December 17, 2002. The spacecraft continued to send valuable scientific data until the end of its mission, sending a total of about 14,000 images and 30 gigabytes of data back to Earth. In all, *Galileo* orbited Jupiter 34 times and traveled a total of 2.9 billion miles (4.6 billion kilometers).

## How did the *Galileo* mission end?

After all its challenges and difficulties, the *Galileo* spacecraft was still working fine, with the exception of the failure of its high-gain antenna. By 2003 it had very nearly used up its propellant. Rather than letting *Galileo* travel in an uncontrolled way, and possibly crash into one of Jupiter's moons by accident, NASA flight controllers decided to end the mission by flying the spacecraft intentionally into Jupiter's atmosphere. So, on September 21, 2003, scientists took one final chance to study the largest planet in our solar system. As *Galileo* went down into the atmosphere and burned up, its instruments recorded the atmospheric and magnetospheric conditions of Jupiter closer-up and more precisely than any previous measurements ever had.

## What is the **Cassini-Huygens** mission?

The Cassini-Huygens mission is a multi-billion dollar international scientific collaboration to study Saturn and its environment—in particular, Saturn's largest moon, Titan. NASA, ESA (the European Space Agency), and ASI (the Italian Space Agency) collaborated on this powerfully equipped exploration mission, which consisted of the *Cassini* orbiter that also carried and supported the *Huygens* lander mission to Titan. Along the way, *Cassini* also flew by Jupiter. With budget constraints and philosophical changes in the world's space agencies, Cassini-Huygens

285

The *Cassini* probe undergoes thermal testing at the Jet Propulsion Laboratory in Pasadena, California. (*NASA*)

is probably the largest and most expensive planetary exploration mission for the foreseeable future.

## How was the *Cassini* spacecraft configured?

*Cassini* is a canister-shaped payload about 22 feet (6.8 meters) long and 13 feet (4 meters) across. A large umbrella-shaped antenna mounted at one end of the spacecraft is its widest feature. Its long radar boom stretches about 35 feet (11 meters) sideways out the side of the spacecraft. Together with the *Huygens* lander, the spacecraft weighed two and a half metric tons, and carried another three tons of fuel at launch. *Cassini* has twelve scientific instruments aboard, and *Huygens* has six more.

## When did *Cassini* launch, and when did it arrive at Saturn?

*Cassini-Huygens* was launched on a Titan IVB/Centaur rocket from Cape Canaveral, Florida, on October 15, 1997. Its flight path included two flybys of Venus, one flyby of Earth on August 18, 1999, and one flyby of Jupiter on December 30, 2000. All four flybys were used as gravitational slingshots to get Cassini-Huygens to Saturn. After nearly seven years in flight, the spacecraft finally arrived at its orbit around Saturn on July 1, 2004.

## What did *Cassini* do while passing Jupiter?

*Cassini*'s long flight to Saturn meant that it was possible—and important—to make scientific observations with the spacecraft while it was en route. From October 2000 through March 2001, *Cassini* made an intensive study of Jupiter as it flew by, taking thousands of pictures and making key measurements in conjuction with the *Galileo* spacecraft. Among its many discoveries, *Cassini* found persistent weather patterns near Jupiter's polar regions; and *Cassini*'s map of Jupiter's magnetic field showed that Jupiter's magnetosphere was lopsided, rather than smooth and round, and had several "holes" where electrically charged particles could "leak" through in huge streams.

## How did *Cassini* enter orbit around Saturn?

When *Cassini* reached Saturn, it fired its rockets for 97 minutes, and used Saturn's own gravity to slow it down. The spacecraft's riskiest moments came when it crossed through the plane of Saturn's rings; the orbital insertion trajectory had

> ### How was *Cassini* used to test Einstein's theory of relativity?
>
> In 2003 astronomers used *Cassini* to test the veracity of Albert Einstein's general theory of relativity with unprecedented accuracy. By comparing changes in the delay time of radio signals to and from *Cassini* as the line of sight varied in distance from the Sun, they were able to see how much the Sun's gravity deflected the signals and thus curved spacetime. The experiment confirmed general relativity to an accuracy of about 0.002 percent—about a hundred times more precise than any previous measurements.

been carefully planned so that *Cassini* would go through the ring plane through a gap, but even one collision with a sizable piece of ring material would have ended the mission. Happily, the orbital insertion was successful. *Cassini* went on to orbit Saturn in a complex, butterfly-shaped (or "Spirograph") pattern, zipping in close and then far away from Saturn in highly elliptical loops, in order to gather close-up data about both the planet itself and its fascinating system of rings and moons.

## What has *Cassini* discovered so far about Saturn?

Among the many discoveries made about Saturn during this mission, *Cassini* has found tremendous dynamic activity inside Saturn's thick atmosphere, including thunderstorms 10,000 times more powerful than anything on Earth that form in huge, deep atmospheric columns about as large as our entire planet. Recently, an Earth-like hurricane—the first to be seen anywhere other than on Earth—was found near Saturn's south pole. It was 5,000 miles (8,000 kilometers) across and packed winds of up to 350 miles (550 kilometers) per hour. *Cassini* has also measured substantial changes in the nature of Saturn's atmosphere since the *Voyager* spacecraft first made measurements in the 1980s. This means that the planet is by no means a static system and continues to change and evolve.

## What has *Cassini* discovered so far about Saturn's moons?

*Cassini* has discovered a number of new moons around Saturn, including several near and even nestled among Saturn's rings. It has also taken spectacular data of many of Saturn's known moons, including Titan, Rhea, Dione, Thetys, Hyperion, and Enceladus.

## What has *Cassini* discovered so far about Saturn's rings?

*Cassini* has taken the most detailed pictures ever of Saturn's rings, and the structures that make the rings so large and complex, yet stable and beautiful. *Cassini*'s discoveries include new ringlets, new moons near the rings, a tiny moon stealing particles from the F ring, another moon (Enceladus) adding particles to the E ring, and features resembling waves, wakes, braids, straw, and rope.

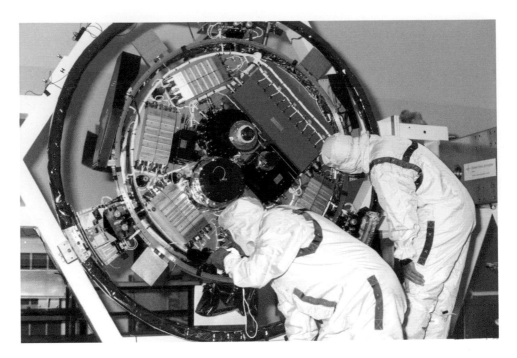

Jet Propulsion Laboratory technicians work on the *Huygens* probe to prepare damage done to its thermal insulation during testing. (*NASA*)

### How was the *Huygens* probe **configured**?

The *Huygens* probe was mounted on the bottom of the *Cassini* orbiter, on the side opposite the high-gain antenna. It was a 700 pound (320 kilogram), four-foot-wide, saucer-shaped vehicle, equipped with a multiple-parachute landing system and six scientific instruments.

### When and how did the *Huygens* probe **land on Titan**?

After riding with the *Cassini* orbiter for more than seven years, *Huygens* separated from *Cassini* on December 25, 2004. It cruised for 2.5 million miles (4 million kilometers) on its own, and entered Titan's atmosphere on January 14, 2005. In an example of wise scientific double-checking, the separation and insertion date were changed years after the mission was launched because it was discovered that a flaw in the computer software would have caused all of the data transmitted from *Huygens* to be lost. The new insertion date, which was selected so that the relative motion between *Cassini, Huygens,* and Earth would allow transmissions to be successfully received, was a month later than originally planned.

*Huygens* hit Titan's atmosphere traveling more than 12,000 miles (19,300 kilometers) per hour. A series of parachute deployments slowed the probe down to less than 200 miles per hour. At an altitude of 75 miles, a final parachute was deployed that slowed *Huygens* further; after more than two hours of falling, *Huygens* landed on the surface of Titan at a speed of just 10 miles per hour.

> ## What is the plan for *Cassini* as the mission continues?
>
> As long as the spacecraft survives, *Cassini* will continue to orbit Saturn and make numerous flybys of rings and moons, even after its primary science mission ends. Like *Galileo* around Jupiter, scientists do not want to contaminate possible ecosystems or pre-biological environments, so it is possible, when the mission is finally over, that flight controllers will crash the spacecraft into an outer moon of Saturn, where it will cause no ecological harm.

## What did the *Huygens* probe reveal about Titan?

*Huygens* sent back 350 images and a wealth of radiometric and meteorological data on Titan. It showed that Titan's atmosphere contains a number of chemicals based on carbon and hydrogen—basic building blocks for more complex organic molecules. It has strong winds, vigorous weather and storm activity, and thunder and lightning. There are clouds and rain—not of water, but of liquid hydrocarbons like natural gas.

*Huygens*'s cameras revealed an amazing variety of geological history on Titan, including free-flowing liquid hydrocarbons on the surface. When *Huygens* landed on the surface of Titan, it hit a thin, brittle crust. Underneath that broken surface was a sandy, swampy substance, which released wisps of methane gas when the probe's impact heated it up. The temperature at ground level was –290 degrees Fahrenheit (–180 degrees Celsius), and the soil consisted mostly of dirty water ice and methane/ethane ice. Pictures of the ground around the landing site showed a surface that looked like a dry riverbed, strewn with smoothed rocks and pebbles.

## What is the *New Horizons* mission?

The *New Horizons* mission is a space exploration probe that is traveling toward the Kuiper Belt. It will fly by Pluto and take the first close look at the dwarf planet, and hopefully other Kuiper Belt Objects as well. Before it received its current name, the *New Horizons* mission was called the *Pluto-Kuiper Express;* before that, it was simply called *Pluto Express.*

## When was *New Horizons* launched and when will it arrive at Pluto?

*New Horizons* was launched on January 19, 2006, on an Atlas V-551 rocket from Cape Canaveral, Florida. Since Pluto is so far away, and time is of the essence to get there, the mission will not have several gravity assists in a long, winding journey to the Kuiper Belt. Instead, the spacecraft was launched directly into an escape trajectory of 36,000 miles (58,000 kilometers) per hour—the highest-speed launch of any space probe—and headed on a beeline for Jupiter. It made a flyby of Jupiter on February 28, 2007, and used the planet in a gravitational slingshot to fly onward toward Pluto. *New Horizons* is expected to fly by Pluto on July 14, 2015.

## What are some of the **instruments carried** by the *New Horizons* spacecraft?

*New Horizons* has seven scientific instruments onboard, one of which (the Student Dust Counter) was built and will be operated by students as the spacecraft travels across the solar system. There is also REX, the Radio Science Experiment; SWAP, which will measure the Solar Wind Around Pluto; PEPSSI, the Pluto Energetic Particle Spectrometer Science Investigation; LORRI, the Long Range Reconnaissance Imager; and a pair of imaging spectrographs.

## What **scientific discoveries** has *New Horizons* already made?

During its rapid flyby of Jupiter, *New Horizons* explored details of the Jovian system that had not been seen before—including lightning near Jupiter's poles, the creation and destruction of ammonia clouds in Jupiter's atmosphere, boulder-sized clumps of rock and ice in Jupiter's rings, the path of charged particles in the tail of Jupiter's magnetosphere, and the internal structure of volcanic eruptions on Jupiter's moon Io.

# EXPLORING ASTEROIDS AND COMETS

## What were some **early efforts** to **explore asteroids** and **comets** with spacecraft?

The first flybys of asteroids were achieved by the *Galileo* spacecraft on its way to Jupiter. In October 1991, the spacecraft flew by Gaspra, and in August 1993 it flew by Ida. These flybys provided the first close images of asteroids; showed that asteroid surfaces are actually quite interesting; and revealed that asteroids could have moons, too—Ida had the asteroid Dactyl for a moon. These flybys helped spur interest in asteroid science, and led to several significant future missions.

The first spacecraft to study comets as their primary science mission were sent toward Comet Halley. In the worldwide anticipation of Halley's return to the inner solar system in 1986, a number of nations worked together to send spacecraft to study the comet and its tail. Two Japanese spacecraft, *Suisei* and *Sakigake,* and two Soviet spacecraft, *Vega 1* and *Vega 2,* had close flybys with Comet Halley in 1986. NASA and the European Space Agency had launched the ISEE-3 satellite on August

12, 1978, and after its original mission was over it was renamed the International Cometary Explorer (ICE) and redirected to observe Comet Giacobini-Zinner in 1985 and Halley in 1986. ICE had no cameras, however. The most significant spacecraft to observe Halley was the *Giotto* mission.

## What was the *Giotto* mission?

The *Giotto* mission was launched by the European Space Agency on July 2, 1985, on an Ariane 1 rocket from Kourou, French Guiana. With information from ICE, *Suisei, Sakigake, Vega 1,* and *Vega 2,* flight engineers were able to get *Giotto* within 370 miles (600 kilometers) of Comet Halley's nucleus on March 13, 1986. Despite suffering damage from several impacts of cometary particles, *Giotto* was able to take spectacular close-up pictures, launching the serious interplanetary study of comets.

*Giotto* was not finished, though. In 1990 ESA flight controllers turned the spacecraft on again from its hibernation mode—the first successful spacecraft restart of its kind—after four years, and redirected it toward Comet Grigg-Skjellerup. It successfully made a flyby on July 10, 1992, coming within just 125 miles (200 kilometers) of the comet's nucleus. Although it could take no pictures—its camera was damaged beyond repair during the Halley encounter—*Giotto* gathered other valuable data. It became the first spacecraft to fly by two cometary nuclei.

## What was the *NEAR-Shoemaker* mission?

The *Near-Earth Asteroid Rendezvous* (NEAR) mission was the first spacecraft sent specifically to explore and orbit an asteroid. The target of choice was 433 Eros, a yam-shaped asteroid whose orbit comes close to that of Earth's. NEAR was launched on February 17, 1996, on a Delta II rocket; after its successful encounter with Eros, the mission was renamed NEAR-Shoemaker, in honor of the pioneering planetary scientist Eugene Shoemaker (1928–1987).

## How was the *NEAR* spacecraft **configured**?

NEAR was shaped like an octagonal prism, about 6 feet (1.7 meters) on a side, with four solar panels and a high-gain radio antenna also about 5 feet (1.5 meters) long. Its scientific payload included an X-ray/gamma-ray spectrometer, a near-infrared imaging spectrograph, a multispectral camera fitted with a CCD imaging detector, a laser altimeter, a radio science experiment, and a magnetometer.

## How did the *NEAR* spacecraft get to **asteroid 433 Eros**?

As NEAR flew toward Eros, it first flew by the asteroid 253 Mathilde on June 7, 1997, and then by Earth on January 23, 1998. Originally, NEAR was to arrive at the asteroid Eros in January 1999, and orbit the asteroid for one year. Unfortunately, just a few weeks before the scheduled rendezvous, an improper firing of the spacecraft's engine put the mission in jeopardy. Instead, scientists flew by Eros on December 23, 1998, and spent more than a year getting the spacecraft back into position for an orbital insertion. On February 14, 2000, NEAR was able to connect with Eros and began to orbit the asteroid.

### What did the *NEAR* spacecraft **reveal** about asteroid 433 **Eros**?

433 Eros is a stony asteroid, shaped roughly like a sweet potato 20 miles (33 kilometers) long and 8 miles (13 kilometers) wide. *NEAR-Shoemaker*'s close-up pictures of Eros showed that it was far more than just a chunk of rock in space. Even though it is a small object, it has had an eventful geological history. A single significant collision with another body about one billion years ago created a single crater, and the ejected material that landed back on the asteroid comprises all of the rocks and dust on the surface of Eros. The collision also sent seismic shockwaves through the entire asteroid, probably changing the shape of Eros and affecting all of the other craters and surface material that were there at the time. Eros has about the same density as Earth's crust—2.4 times that of water—and tumbles through space as it orbits the Sun, making a complete orbit every 643 days and one complete rotation every 5 hours, 16 minutes.

### How did the *NEAR-Shoemaker* mission **end**?

On February 12, 2001, flight engineers gently flew the *NEAR-Shoemaker* spacecraft onto the surface of 433 Eros. It landed at a speed of about three miles per hour—about fast walking speed—and, to the scientists' delight, the spacecraft survived with only minor damage, even though it was never designed to land. After gathering data at the surface for a few more weeks, *NEAR-Shoemaker* was shut down on February 28, 2001.

### What was the *Deep Impact* mission?

*Deep Impact* was a mission to hit a comet with a hard, dense object at high speed, and then take pictures and gather other data of the impact site and the ejected material. The reason for such a study was to see what the interior of a comet—the oldest unaltered material in the solar system—could reveal about the origin of the

The *Deep Impact* mission taught astronomers about the materials that make up a comet, including clay, carbonates, crystallized silicates, polycyclic aromatic hydrocarbons, iron compounds, and even bits of the reddish-brown gem spinel. *(NASA/JPL-Caltech/R. Hurt (SSC))*

planets, and to learn how to deal with a future comet that might be on a collision course with Earth. The space explorer mission was combined with a concentrated effort of ground-based and space-based telescopes to study the comet and observe the impact and its aftermath.

### How was the *Deep Impact* spacecraft **configured**?

The *Deep Impact* spacecraft had two parts: the flyby craft, which is about 10 feet (3 meters) long, 6 feet (1.8 meters) wide, and 8 feet (2.4 meters) high and outfitted with sensitive scientific instruments; and the impactor, which was a 820-pound (370-kilogram), metal (mostly copper) box about the size of a washing machine and outfitted with a camera and a small thruster.

### **How** and when did *Deep Impact* **collide** with its target **comet**?

*Deep Impact* was launched on January 12, 2005, by NASA on a Delta II rocket, toward the comet P/Tempel 1 (also simply called Comet Tempel 1). On July 3, the impactor separated from the flyby craft and guided itself into the path of the oncoming comet. The next day, as scientists watched with their instruments and the public watched on the Internet, Comet Tempel 1 crashed into the impactor at more than 23,000 miles (37,000 kilometers) per hour.

### What **happened** when *Deep Impact* collided with **Comet Tempel 1**?

The cometary material that the *Deep Impact* collision kicked up was so copious and reflective that cameras and instruments could not see the crater itself. But the scientific return of the impact was enormous. For the first time, astronomers were able to

293

study the unaltered ice and dust that existed in the solar system more than four billion years ago. The result also showed how soft—and powdery!—comets can be; this will be important if humans someday need to move a comet hurtling toward Earth, because using the wrong technology on the wrong material would not do the job.

## What was the *Stardust* mission?

The *Stardust* mission was launched on a Delta II rocket from Cape Canaveral, Florida, on February 7, 1999. Its destination was Comet Wild 2, and its objective was to capture grain-sized pieces of the comet's coma, then return toward Earth. Both on the way to the comet and back, *Stardust* would capture interplanetary dust grains that it encountered during its journey. The dust grains would then be safely returned to Earth's surface for study, and the spacecraft would fly on past our planet.

## Why is it important to explore **interplanetary space dust**?

The solar system is far more empty than it is full. Even though people pay the most attention to the largest objects in it, the rest of the system should not be neglected. All larger solar system objects were built by smaller ones, and everything started out as dust. So the tiniest particles floating in the solar system may contain some of the best clues about the origin of the solar system and its contents, as well as the solar system environment we live in today. They may even contain clues about other stars in our galaxy and their origins.

## How was the *Stardust* spacecraft **configured**?

*Stardust* is about the size and shape of a large refrigerator. Along with cameras and other scientific instruments, a special return capsule was mounted where the cometary and interplanetary particles would be captured, stored, and then returned safely to Earth.

## **How** did the *Stardust* spacecraft **capture** cometary and interplanetary **particles**?

When *Stardust* was in particle-capturing mode, the return capsule was opened and an arm was extended away from the spacecraft. On this arm, which looked like a

## What is Aerogel?

**A**erogel, sometimes nicknamed "frozen smoke," is a type of solid, translucent foam material that is almost completely (about 99.8 percent) air. It can be made out of different substances, such as silica, carbon, or alumina (aluminum oxide). It is the lightest solid substance manufactured by humans, yet it has remarkable thermal insulating properties and structural strength. Spacecraft designers often use Aerogel to insulate their payloads (such as the Mars Exploration Rovers). Aerogel was the perfect substance to use in the *Stardust* mission to catch speeding cometary and interplanetary particles without causing them to be destroyed, either by the impact or the heat caused by friction.

tennis racket or catcher's mitt, were flat mounted trays of Aerogel—a superlight, superstrong substance. When the particles struck the Aerogel, they would suddenly slow down from over 10,000 miles per hour to zero in a split second, embedding into the matrix without breaking or melting. When the particle collection was over, the arm retracted and the Aerogel was safely stored in the return capsule.

## How did the *Stardust* mission return cometary and interplanetary particles to Earth?

With its collectors open, the *Stardust* spacecraft flew by Comet Wild 2 on January 2, 2004. On the morning of January 15, 2006, the return capsule was released toward Earth as the spacecraft skimmed the very top of Earth's atmosphere. The capsule came in on a nearly flat trajectory at some 29,000 miles (46,500 kilometers) per hour—the fastest re-entry ever made by a man-made object. With a series of parachutes slowing it down, the capsule landed safely in the Utah desert with more than a million cometary and interplanetary particles safely embedded in blocks of Aerogel ready for scientific study.

## What is the status of the *Stardust* spacecraft and mission?

Shortly after the successful capsule return, *Stardust* was put in a hibernation mode as scientists pondered what to do next with the still-functioning spacecraft. It was decided to re-purpose the spacecraft on a new extended mission to go to Comet Tempel 1, the target of the *Deep Impact* mission, and gather new pictures and other data of the comet and its new crater. The spacecraft has been re-activated, and is headed for its new destination as the NExT mission (New Exploration of Tempel 1).

# LIFE IN THE UNIVERSE

## LIVING IN SPACE

### Can **humans live** in **space**?

Humans not only can live in space—we already *do* live in space! Since 1971, humans have been maintaining space stations in low Earth orbit, where people can stay in outer space for extended periods of time. Human beings have now lived continuously in space for nearly a decade. They have been doing so for so long, in fact, that most people on Earth do not even give it a second thought any more. The challenge now for humanity is to live in outer space beyond low Earth orbit, such as on the Moon, or Mars, or in interplanetary or interstellar spacecraft.

### What **life support** is necessary for humans **to live** in space?

In space every environmental need for humans to survive—including air to breathe, water to drink, food to eat, and room to move around—must be provided by artificial methods. This means a fully contained life support environment must exist, including everything from light and heat to air recycling and waste removal. Above Earth's atmosphere, it is also essential for any human habitat in space to provide protection against hazards in the space environment, such as excessive radiation, cosmic rays, or meteoroids.

### What **happens** to the **human body** in space?

In orbit or in deep space, humans are weightless; that is, the net force on their bodies from gravity is zero. This is not because they are far away from Earth, but rather their orbital speed and trajectory create acceleration that exactly balances Earth's gravitational acceleration. Since humans evolved in an environment where gravity is not zero, our biological systems react significantly to the zero-gravity or micro-

297

Many adjustments have to be made for humans to live in a zero-gravity environment. Here, space shuttle astronauts Kathryn Sullivan (left) and Sally Ride show the velcro and bungee cord restraints used to keep people from floating away as they sleep. (*NASA*)

gravity environment. Bodily fluids like blood fill the face, puffing up the skin; muscle fibers grow thinner with disuse, causing muscles to weaken and atrophy; the mineral turn-over process slows down in bones, causing a decrease in bone density akin to osteoporosis. Thus, when people are in space for any prolonged period of time they must conduct rigorous physical activity and exercises in order to stay healthy.

## What was **life** like on *Skylab*?

Living conditions on *Skylab* were remarkably comfortable, considering that it was basically a big tin can in space. The living area was quite large, and the sleeping accommodations were private. The kitchen area included a freezer containing 72 different food selections and an oven of sorts. The dining table was placed beside a window so crew members could enjoy a view of space while they ate. *Skylab* even had the first space shower and private toilet. (The toilet employed a seat belt to prevent the user from floating off.)

## How did the astronauts **exercise** while aboard *Skylab*?

To keep the astronauts healthy and combat the atrophy of their physiological systems, *Skylab* had exercise equipment aboard, including a stationary bicycle and a treadmill. An odd consequence of exercising, however, was that sweat floated off the astronauts' bodies in slimy puddles. The person working out had to catch these puddles with a towel so that the moisture would not land on anything that might be damaged by excessive moisture!

Aboard the *Mir* space station, American astronaut Shannon Lucid exercises in order to prevent her bones and muscles from weakening. (*NASA*)

## What was **life** like aboard *Mir*?

Living space on *Mir* consisted of two small sleeping cabins and a common area with dining facilities and exercise equipment. It accommodated three people at a time for indefinitely long stays, and up to six people for short stays of up to a month. Although it was cozy, *Mir* was designed with comfort and privacy in mind, since it was correctly anticipated that crew members would be living on the station for very long periods of time.

## How **long** have humans **lived** on the **International Space Station**?

The ISS has been continuously inhabited by at least two people since November 2, 2000. The plan is for ISS to remain inhabited until at least 2016. Astronauts and cosmonauts from more than a dozen nations have visited ISS, as well as the first-ever "space tourists"—civilians who have paid a fee to ride a rocket up to the station, spend some time onboard doing simple tasks, and then ride back down again. On August 10, 2003, a cosmonaut even got married on the ISS. He exchanged vows with his bride while he was in orbit over New Zealand and she was on the ground in Texas.

## What is the **International Space Station**?

The International Space Station (ISS) is a multinational research vessel that is currently orbiting at an altitude of 210 miles (340 kilometers) above Earth's surface. The ISS project grew out of an agreement between the governments of the United States and Russia, who both wanted to build a permanent human presence in space but lacked the political will and funding to do so separately. With the breakup of the Soviet Union and end of the Cold War in 1991, civilian space projects took a backseat to other funding priorities in both superpowers; this meant that the Americans' *Freedom* and the Russians' *Mir 2* space station programs were at a near-standstill. In 1993 an agreement was reached to build a new, fully international space station to be completed by 2010, a plan that was ultimately palatable to the voters and taxpayers of both nations.

Today, the ISS is a joint project of the space agencies of Russia, Europe, Canada, Japan, Brazil, Italy, and the United States. The first ISS module was launched on November 20, 1998, by a Russian Proton rocket; the second module, called "Node 1," was brought into orbit by the space shuttle *Endeavour.* The fully completed ISS is expected to have 14 pressurized modules in all, contain about 30,000 cubic feet (1,000 cubic meters) of interior space, and have a mass of more than 450 tons.

## How is the **International Space Station configured**?

Even halfway toward completion, the ISS was already the largest space station ever built. It has a long, narrow main truss, with several modules branching horizontally outward, and several sets of solar panels to provide electric power to the station's many systems. The span of the ISS solar panel arrays is about the length of a football field. Major pieces of ISS were originally designed to be parts of other, separate space missions—such as the American space station *Freedom,* the Russian space station *Mir 2,* the European *Columbus* program, and the Japanese *Kibo* experiment module—and adapted for inclusion into ISS.

## What is the **value** of the **International Space Station**?

Critics of the ISS have long argued that the project is, to varying degrees and in various ways, an expensive boondoggle with little bang for the buck. Any scientific

An artist's drawing of the space station *Freedom* with a shuttle preparing to dock. (*NASA*)

return of ISS could be achieved much more cheaply, they argue. Further complaints have been brought to bear against the involvement and administration of so many international partners has led to waste and inefficiency, and the danger and great cost of supporting human life in low-Earth orbit has drained resources away from many other worthy space-based projects.

While these arguments may have substantial merit, another way to look at ISS is not purely through an economic lens, but rather from a more holistic sociological perspective. No space program in history has ever been inexpensive, and all of them have had their share of embarrassments, failures, and tragedies. Even so, human spaceflight and space exploration has led the way for us as a species to reach beyond the limits imposed upon us by Earth's natural boundaries. In some ways, ISS is almost a victim of its own success. Humans have been living on ISS continuously for so long now that space travel from Earth to the station seems routine and commonplace; it fails to capture the excitement and interest of taxpayers and lawmakers. Excluding space shuttle costs, the U.S. government spends about two billion dollars per year on ISS. That is a lot of money, but it turns out to be less than two cents per day per American citizen. The stimulus of ISS to our creativity, imagination, and desire to learn and grow may, ultimately, be well worth this price.

## LIFE ON EARTH AND ON THE MOON

### What makes **Earth so unique** in our solar system?

As far as people know, Earth is the only place in the universe that supports life. Many scientists believe that one day we will find life elsewhere in the cosmos, but even if we do discover other life forms out in the solar system, galaxy, or universe, we should realize that life is precious and we should value the fact that we live on a planet teeming with plant and animal species.

### How is **Earth's atmosphere** important to **life** on Earth?

Very few life forms on Earth can survive for any length of time at all without Earth's atmosphere. We breathe the atmosphere; and it blocks harmful radiation from space. The pressure it provides keeps surface water liquid, and the greenhouse effect it produces keeps us warm.

### Is the **greenhouse effect** a good thing for life on Earth, or bad for the environment?

Like so many things about life on Earth, moderation is the key. Some greenhouse effect is a very good thing for life on Earth. Without any such effect on Earth, the oceans would eventually freeze. If the greenhouse effect increases significantly, however, many living organisms and species, as well as environmental systems that have developed over a long time—including human civilization—will face substantial challenges, and possibly even extinction. In the most extreme case, a runaway greenhouse effect like that on the planet Venus, would cause all life on Earth as we know it to cease to exist.

### Why is the **ozone layer important** to life on Earth?

This ozone layer is important to life on Earth because ozone, which has three oxygen atoms as opposed to two for ordinary oxygen, effectively absorbs energetic ultraviolet rays, which are harmful to plants, animals, and people.

### Why is Earth's **magnetic field important** to life on Earth?

Earth's magnetic field extends out into space, creating a structure called a magnetosphere, which surrounds our planet. When the magnetosphere is hit by charged particles from space, such as from the solar wind or from a coronal mass ejection, it deflects these particles away from Earth's surface, significantly reducing the amount that strikes life forms down on Earth's surface. This protects us from the hazards of being hit by too many such particles.

Ocean tides created by the Moon's gravitational pull provided an opportunity for early life on Earth to make the transition from sea to land in tide pools. (*iStock*)

## Why were **ocean tides** important to the **evolution of life** on Earth?

All animal life on Earth used to live only in the oceans. Scientists think that, for the evolution of land-based life on Earth to occur, it would have been important to have a transitional zone between ocean and land; that is, shorelines that were at times dry, then at times wet, over a long and regular cycle. This way, animals could evolve by slowly adapting to life in drier environments. Over millions of years, these animals could eventually evolve into animals that lived and breathed exclusively on land. Areas with regular, vigorous ocean tides provide just such a transitional zone, becoming wet and dry over and over again every 13 hours or so. Thus, land-based animals like humans may well have had their evolutionary start in the tidal basins and tidepools of ancient continental coastal areas. Without the Moon, such tides would not be present, so the Moon has also proved to be vital to evolution on Earth.

## Could **comets** have been the **source for water** and **life** on Earth?

Since comets contain huge amounts of ice and rock, astronomers have long speculated that comets colliding with Earth may have deposited substantial amounts of water onto Earth's surface early in our planet's history. Recently, additional hypotheses have been proposed that biological ingredients for life—such as complex protein and DNA molecules—may have been formed elsewhere in the solar system or the galaxy; they may have been frozen into cometary ice and then carried down to Earth's surface billions of years ago, seeding our planet with the biological precursors for life. The latest research suggests, however, that while some water

## How does Jupiter protect life on Earth?

Even though Jupiter is on average half a billion miles from Earth, its strong gravity may have played a profound role in the development of life on Earth. Jupiter pulls all kinds of matter toward itself with its gravity, including comets and asteroids. Had Earth been bombarded with too many comets and asteroids over the past four billion years, life on Earth might not have had a chance to develop and evolve as it has. Jupiter has acted like a gravitational shield, absorbing large numbers of cosmic projectiles that otherwise might have caused serious blows to life on Earth.

may well have been brought to Earth by extraterrestrial objects, complex organic molecules probably break down too quickly when exposed to the extreme cold and radiation environment of interplanetary space to survive a multi-million-year ride embedded in cometary ice.

### Is there **liquid water** on the **Moon**?

No liquid water exists on the Moon's surface. This is because there is no atmosphere on the Moon, and without atmospheric pressure water cannot remain liquid. There is also no evidence that there is liquid water under the surface of the Moon.

### Is there **ice** on the **Moon**?

There is evidence that water ice crystals do exist on the surface of the Moon. In 1994 the *Clementine* lunar probe made radar measurements of the Moon's south pole that indicated the possible presence of frozen water mixed into the lunar soil and rocks. The particular location where the measurement was made is about the size of four football fields, and reaches a depth of about 16 feet (5 meters). The ice is thought to be inside a deep crater, and may have been deposited there when comets, which are comprised mostly of water ice, crashed into the Moon's surface. Deep inside the craters, where sunlight does not shine, the ice may have accumulated without being melted by the Sun's heat.

# LIFE IN OUR SOLAR SYSTEM

### What did the *Cassini* discover on Enceladus?

One fascinating discovery is that Enceladus contains liquid water, spouting it into deep space in huge geysers. These frozen water droplets comprise a large part of Saturn's faintest and most distant ring, the E ring, and suggests that Enceladus may be a place to look for extraterrestrial life.

## Where in the **solar system** does **liquid water exist**?

Liquid water is known to exist in abundance on Earth's surface. Detailed studies of Mars over the past decade have shown tantalizing evidence that liquid water exists there underground, and occasionally spurts out through canyon walls and other geologic events on the Martian surface. Studies made with the *Galileo* spacecraft show that Europa and probably Ganymede, two of Jupiter's moons, probably contain liquid water deep beneath their surfaces; and studies done with the *Cassini* spacecraft show that Saturn's moon Enceladus blows geysers of liquid water into deep space, through fissures on its icy surface.

## Where in the **solar system** do the **chemicals** to **support life** exist?

Almost every solar system body has embedded within it the chemical elements necessary for the formation of life as we know it. This is probably because those elements—chiefly hydrogen, oxygen, carbon, and nitrogen—are among the most common elements in the universe. Places especially rich in these chemicals include the atmospheres of the gas giant planets; the surfaces of Earth, Mars, and Titan; and possibly deep underground on Europa and Ganymede.

## Where in the solar system does **steady energy exist** to **support life**?

The most plentiful source of steady energy in our solar system comes from the Sun. In a particular zone around the Sun—not too far away, but not too close—solar radiation is intense enough to melt ice into liquid water, but not so harsh as to vaporize the liquid water into steam. Earth happens to be in the Sun's habitable zone.

Interestingly, under the surface of many solar system bodies, steady energy may also come from deep within the core. If tidal interaction is present, energy may constantly be flowing throughout the space body; and if mass differentiation is still going on, where dense metallic material sinks slowly through the bodies' lighter rocky or gaseous layers, the gravitational potential energy released by that process can be both gentle and persistent over very long timescales. Solar system bodies other than Earth where underground energy sources may be enough to support life include Mars, Europa, and Ganymede.

## What did the *Huygens* probe reveal about the possibility of **life** on Saturn's moon **Titan**?

Scientists have long speculated that Titan might have the chemical ingredients for the development of life, and they wondered if *Huygens* might find living things on the surface. As it turned out, though, *Huygens* found nothing alive, but it did provide evidence to confirm some important hypotheses about life-like indicators on Titan.

For example, astronomers wanted to explain how methane could persist on Titan, even though the Sun's ultraviolet light should theoretically destroy all free methane gas. On Earth, the methane gas in the atmosphere is replenished by living organisms; on Titan, though, it is too cold for life to survive. Thanks to *Huygens* data matched with theoretical simulations, planetary scientists now realize that geological process-

305

es—venting and volcanoes—fill the Titanian environment with methane, much the way water vapor was deposited into Earth's atmosphere billions of years ago.

Even though *Huygens* found no life on Titan, it confirmed that Titan has all the essential chemical ingredients to foster biological processes like we have on Earth. Furthermore, with the discoveries of liquid methane lakes, rivers, streams, and seas on Titan, and its dynamic and ever-changing environment, scientists have lots of new data to explore in their continuing contemplation of the search for life on other worlds.

## Could there be **life** on Jupiter's moon **Europa**?

Studies suggest that, miles beneath the solid surface of Europa, there is a vast underground ocean of liquid water. Great controversy exists about whether or not that underground ocean, like oceans here on Earth, could be an ecosystem where life as we know it may thrive.

## What was the **primary mission** of the **Viking program**?

In the 1970s, there was still uncertainty about the existence of life on the surface of Mars. Although the current conditions had been shown to be inhospitable to life as we know it on Earth, data from the Soviet Mars series and U.S. Mariner series of space probes suggested that the cold, dry periods of the Martian present may have alternated with warm, moist periods in the Martian past, with each cycle lasting perhaps 50,000 years. This raised the possibility that life forms may have evolved on the surface of Mars that would lie dormant during the hostile climatic periods and then reactivate when the climate was more hospitable. Thus, the primary mission of the Viking probes was to examine Mars thoroughly for any signs of life, dormant or otherwise.

## What was the evidence for **fossilized life** in the **Martian meteorite ALH84001**?

Dozens of scientific research groups are studying this Martian meteorite—and pieces of it—to figure out if there indeed is fossilized life embedded within it. The

> ## How were pulsars related to "little green men"?
>
> The first pulsar's radio signals, discovered by Jocelyn Susan Bell Burnell (1943–) and Antony Hewish (1924–) in the 1960s, came regularly at 1.337-second intervals. The periodicity of the pulses was so perfectly regular that it was hard to conceive at that time of any naturally occurring phenomenon that would cause it. Here on Earth, only living things and man-made machines can make such perfectly regular, periodic phenomena occur. So Bell Burnell and Hewish wondered whether the pulses originated from extraterrestrial life. They nicknamed their original pulsar "LGM," an abbreviation for "Little Green Men." It turned out the explanation was just as extraordinary: a rapidly spinning, electromagnetically charged neutron star.

primary evidence consists of tiny, sausage-shaped markings in the meteorite less than a billionth of an inch long. These imprints resemble some kinds of fossilized bacteria found in rocks on Earth. Some of them are assembled in long chains of iron-rich magnetite crystals, which on Earth are usually produced by biological processes in microorganisms.

# SEARCHING FOR INTELLIGENT LIFE

## How do scientists look for **intelligent extraterrestrial life**?

The search for extraterrestrial intelligence, or SETI, is a fascinating enterprise that has captured the imaginations of creative thinkers for generations. For a long time, SETI was not considered mainstream science by most people. Today, though much of the worldwide conversation on SETI remains in the realm of speculation and pseudoscience, scientifically legitimate and credible SETI efforts are indeed going on.

Modern search missions have most commonly employed radio telescopes that target nearby stars similar to the Sun. They act as antennae, listening for communications signals created by alien civilizations that are sent out from their home planets, either accidentally or intentionally.

## What is **challenging** about **listening** for **alien radio signals**?

The success of this kind of search depends not only on the existence of intelligent extraterrestrial life, but also on life intelligent enough to figure out how to send such signals. Furthermore, radio signals weaken fairly quickly. After traveling even a few dozen light-years, the interstellar medium can scatter and muffle such signals significantly, so much so that even Earth's largest radio telescopes would not be able to detect them.

## Who **pioneered** the scientific **search for extraterrestrial intelligence**?

The American astronomer Frank Drake (1930–) is widely considered to be the first person to pursue the scientific search for extraterrestrial intelligence. Drake grew up in Chicago and earned his degrees at Cornell University and at Harvard. In 1960 he conducted his first search for extraterrestrial intelligence by using radio telescopes—an endeavor called Project Ozma. He co-organized the first scientific conference on SETI, was instrumental in creating the SETI Institute, and was the first person to conceive what is now known as the Drake Equation.

When asked what motivated his interest in SETI, Drake answered, "I'm just curious. I like to explore and find out what things exist. And as far as I know, the most fascinating, interesting thing you could find in the universe is not another kind of star or galaxy or something, but another kind of life."

## What is the **Drake Equation**?

The Drake Equation (sometimes called the Green Bank Equation), named for SETI pioneer Frank Drake (1930–), is a mathematical expression that encapsulates the conceptual framework of SETI. According to this equation, the number of alien civilizations in our Milky Way galaxy with which humans could communicate is the product of seven factors: (1) the average rate of star formation in the Milky Way; (2) the fraction of those stars that have planets; (3) the average number of habitable planets that each planet-hosting star has orbiting around it; (4) the fraction of habitable planets that actually have life forms on them; (5) the fraction of life-bearing planets that develop civilizations of intelligent life; (6) the fraction of civilizations that produce detectable signs of their existence, such as radio waves or atmospheric changes; and (7) the length of time that such civilizations emit those detectable signs.

The Drake Equation is a useful way to think about SETI scientifically. Each of the seven factors can be studied with the methods of science. Unfortunately, at this point in human history, we simply do not have enough information to know the actual value of most of the seven factors with any degree of accuracy. It is a sure bet, though, that astronomers will keep trying to find them out. Already, astronomers think that one Sun-like star is formed in the Milky Way every few years or so; this star formation rate is still only an estimate.

### Is there a possibility that **intelligent extraterrestrial life might find us** first?

Humans have sent into space enough evidence of our existence that intelligent extraterrestrial life may indeed find us before we find it. In 1974 astronomers used the Arecibo Radio Telescope in Puerto Rico to beam a short radio message toward Messier 13, a globular cluster of hundreds of thousands of stars about 25,000 light-years away from Earth. Radio and television signals have been emanating from broadcast towers for about half a century. There are also physical objects that have gone beyond the orbits of the planets in our solar system, such as *Pioneer 10*, *Pioneer 11*, *Voyager 1*, and *Voyager 2*. Onboard those spacecraft, astronomers have installed pictures and audio recordings about our solar system, our planet, and ourselves.

### What was the purpose of putting **gold plaques** on *Pioneer 10* and *Pioneer 11*?

A gold plaque was carried on both *Pioneer 10* and *Pioneer 11*. The plaques were engraved with information about Earth and humans, just in case they ever encounter intelligent life as they journey through deep space.

### What is the **Golden Record** aboard the **Voyager** spacecraft?

Each of the two Voyager spacecraft, which are now well beyond the orbit of Neptune, carries onboard a gold-plated phonograph record that bears pictographs showing how to play them using simple (by human standards) electrical technology. Each "Golden Record" contains about two hours of sounds that can be heard on Earth, such as falling rain, thunder, bird and animal calls, human speech, and all kinds of music. If an intelligent species found one of the Voyager probes, it might be possible for them to learn something about humans and about Earth—a distant, friendly greeting from us to them.

The Voyager spacecraft carry gold records with sound recordings from Earth that, one day, might be listened to by alien ears. (*NASA*)

### What might be the **strongest argument against** the belief that **intelligent extraterrestrial life** exists?

The Italian physicist Enrico Fermi (1901–1954) was once asked if intelligent extraterrestrial life exists. Fermi replied, "Where are they?" The so-called Fermi Paradox about extraterrestrial intelligence can be simply summarized. If technological advancement on Earth follows its current trajectory, in a few centuries or millennia we will be an interstellar space-faring species. After that point in human history, even if it took our spaceships a century to get to the nearest stars, humans could populate the entire galaxy in about ten million years. Since the Milky Way has been forming stars for about ten billion years—a thousand times that long—a human-like civilization would develop into such an advanced civilization very quickly compared to the age of the galaxy. If even one such advanced civilization existed in our galaxy, then evidence of their existence should be abundant in our astronomical observations. Since such evidence has not been observed, it is reasonable to think that such a civilization does not exist.

### What is the **strongest argument in favor** of the existence of **intelligent extraterrestrial life**?

Currently, the strongest argument that extraterrestrial life exists is as follows: (1) there are so many planets in the universe that some of them must have an environment similar to that of Earth; (2) in every environment on Earth that has been examined, life has been found; (3) the laws of nature are universal, so Earth-like planets should be able to support life as we know it in the same way that Earth itself does. With this line of reasoning, it seems all but certain that life must exist somewhere in the cosmos other than just here on our planet.

### What is one **argument** that **intelligent life existed**, but that it **failed to contact us**?

A counter-argument to those who believe no other intelligent life exists in the galaxy is as follows. Any intelligent civilization, like our own human civilization, would be tempted to use any new technology for weaponry against its enemies. It is therefore possible that every intelligent civilization destroys itself before it can expand beyond its own solar system. Given humanity's own self-destructive technologies, such as nuclear and biological weapons, the history of our own species has yet to show whether or not that hypothesis could be true.

# EXOPLANETS

### What is an **exoplanet**?

An exoplanet, also known as an extrasolar planet, is a planet that is not in our solar system. The first confirmed exoplanet was discovered in the late 1990s; since then, more than 200 exoplanets have been discovered, and the number is increasing at the rate of more than a dozen new exoplanets per year.

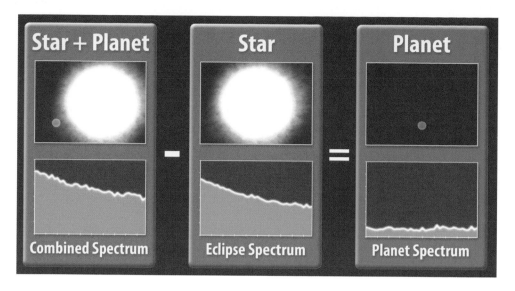

By analyzing changes in spectra around a star, astronomers can detect whether a Jupiter-sized planet is orbiting close by. (*NASA/JPL-Caltech/R. Hurt*)

## How do astronomers **find exoplanets**?

The most common method to date for finding exoplanets is by the Doppler method, which uses the Doppler shift of light. As a planet orbits its star, it moves the center of gravity of the star system back and forth. By measuring that motion in the spectra of stars, it is possible to deduce whether or not a planet is orbiting that star, how massive it is, and what its orbital distance is.

Another method for finding exoplanets is looking for the shadow of a planet moving across our line of sight to its star. This method will find fewer planets, because such eclipsing binary exoplanets are very rare, but when they are found astronomers can learn much more about the exoplanet than with only the Doppler method. Parameters including size, temperature, chemical composition, and atmospheric density are measurable using the shadow method.

Most astronomers feel that the best way to find exoplanets would be to take a direct image of them. Unfortunately, that method is impossible with current astronomical instrumentation because the host stars outshine the planets by such a huge factor that it would be harder than finding a firefly in the beam of a searchlight. Scientists are working diligently, though, to develop the technology that will allow us to compensate for the effects of such a huge contrast level. Perhaps within a few years, we will be able to look at a picture of an exoplanet in a distant star system.

## How can **interferometry** be used to find **extrasolar planets**?

Just as it is possible to use interferometry to obtain very detailed images of objects in space, the interference patterns of light can be analyzed to take very detailed spectra. The resulting measurements of Doppler shifts in the spectra—and hence, the motions of the objects producing the shifts—can be extremely precise. With

311

## What have exoplanetary systems taught us about our own solar system?

The study of exoplanetary systems has rewritten the book on what planetary systems generally are like. Before exoplanets were discovered, astronomers only had our solar system to refer to, so theoretical models all used it as the basic template for all planetary systems. Now, hundreds of exoplanetary systems are known, and none of them are like ours. Even though most scientists think systems like our own will eventually be found, it is quite clear already that a large fraction of planetary systems bear no resemblance to our solar system. Theoretical models of planetary systems and planet formation now include a great deal more variety than just a decade ago. We now understand that even though the solar system follows all the same laws of nature as every other planetary system, it is a surprisingly special place.

current technology, for example, it is actually possible to measure changes of speed as little as a person running or jogging at distances of hundreds of trillions of miles!

As it turns out, this is the level of speed changes that large planets exert on the stars they orbit. If, for example, the gas giant planet Jupiter orbited the Sun at the distance of Mercury, the Sun would wobble back and forth, changing the direction of its motion every few weeks by an amount similar to a running person. By measuring the spectra of nearby Sun-like stars, and using interferometry to detect the minuscule changes in their speed, it is possible to detect and confirm the existence of planets orbiting around them. Hundreds of extrasolar planets have been detected in exactly this way, all of them orbiting faraway stars.

### What are **extrasolar planetary systems** like?

Exoplanetary systems are very different from our solar system—at least, from what astronomers have been able to observe so far. Most of these systems have, for example, huge gas giant planets orbiting closer to their stars than Mercury is orbiting around the Sun. That characteristic alone would destroy every Earth-like planet in the inner parts of those systems.

### Have any **exoplanets** been **discovered** yet that are **like Earth**?

Alas, not yet. Even with the best current technology, astronomers cannot see the planets in exoplanetary systems directly. We do know, however, that all the exoplanetary systems that have been detected so far have at least one large gas giant planet, about the mass of Saturn or greater, orbiting very close to its host star. It is possible that some of these systems have smaller, terrestrial planets, but we can neither see nor detect them right now. Generally speaking, since the laws of physics are the same around our Sun as they are around any other star, it is safe to guess that ter-

An artist's illustration of what the Jupiter-like exoplanet HD 14902 6b might look like. This gas giant orbits close to its sun and has average temperatures of 3,700 degrees Fahrenheit (2,040 degrees Celsius); it is also very hot—and dark—because its atmosphere absorbs most of the energy from its sun. (*NASA/JPL-Caltech/T. Pyle*)

restrial planets might even be common in exoplanetary systems. Until we can observe them directly, though, we will not know for certain.

## What are some of the **difficulties** of finding **Earth-like extrasolar planets**?

It is important to realize that all the exoplanetary systems we have found so far are constrained by the technology we use to find them. We actually cannot detect the motions of Earth-like planets at all beyond our own solar system, and can only confirm the presence of planets much larger and much more massive. Astronomers are also limited by the amount of time they have been able to search for exoplanets. Jupiter takes more than a decade to orbit the Sun once, so scientists have to observe a distant Sun-like star even longer than that to detect a planet with an orbit like Jupiter's. As technology improves, it becomes increasingly likely that we will eventually find a planetary system that is very similar to our own solar system.

## What have **exoplanetary systems** taught us about **our own solar system**?

The study of exoplanetary systems has rewritten the book on what planetary systems generally are like. Before exoplanets were discovered, astronomers only had our solar system to refer to, so theoretical models all used it as the basic template for all planetary systems. Now, hundreds of exoplanetary systems are known, and none of them are like ours. Even though most scientists think systems like our own will eventually be found, it is quite clear already that a large fraction of planetary systems bear no resemblance to our solar system. Theoretical models of planetary systems and planet formation now include a great deal more variety than just a decade ago. We now

313

understand that even though the solar system follows all the same laws of nature as every other planetary system, it is a surprisingly special place.

# LIFE ON EXOPLANETS

### What is meant by **"life in the universe"**?

The concept of life in the universe represents a number of ideas that have long stoked the fires of human imagination. These ideas include life on Earth traveling out into the universe—that is, to outer space; life from Earth, especially humans, living elsewhere in the solar system or the universe; and life *not* from Earth—that is, extraterrestrial life—and our search for it.

Although people have pondered these ideas since time immemorial, it has only been recently—within a human lifetime—that humanity has made significant strides in these endeavors. Since the 1950s, we have sent rockets and satellites into space. Since the 1960s, we have sent humans into space and returned them safely to Earth. Since the 1970s, we have sent humans to live in space for weeks, months, and even years at a time. Since the 1980s, scientifically significant searches for extraterrestrial life have been made. And since the 1990s, astronomers have discovered hundreds of planets orbiting stars other than the Sun.

### What is life **"as we know it"**?

Ironically, though astronomers search for life beyond Earth, biologists here on Earth have still not conclusively determined what constitutes life as we know it. The basic definition of a living thing is something that begins an active existence (is born), changes and matures over time (grows), replicates itself through a well-ordered process (reproduces), and then ends its existence (dies). On Earth, all things that go through these steps achieve them through the complex interactions of very large molecules such as ribonucleic acid (RNA) and deoxyribonucleic acid (DNA). Some terrestrial things, however, go through all four of those steps, at least by some definitions, but may or may not be alive; certain viruses, for example, defy

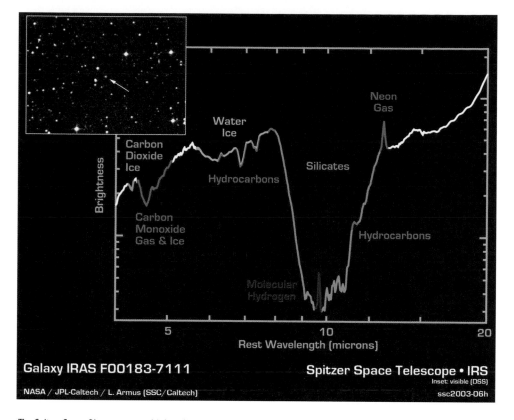

Galaxy IRAS F00183-7111

Spitzer Space Telescope • IRS
Inset: visible (DSS)

NASA / JPL-Caltech / L. Armus (SSC/Caltech)

ssc2003-06h

The Spitzer Space Observatory used infrared spectroscopy to analyze the distant galaxy IRAS F00183-7111, where water and organic compounds were successfully detected. (*NASA/JPL-Caltech/L. Armus, H. Kline, Digital Sky Survey*)

easy classification. In the cosmic context, the line between living and non-living things may be even blurrier; stars, after all, are born, grow, reproduce, and die—all after a fashion, at least. So, are stars alive?

## What do **astronomers look for** when they search for **extraterrestrial life**?

It is still beyond current technology to look for individual living things. When astronomers search for extraterrestrial life, therefore, the targets are ecosystems—environments on other worlds that could harbor life. As far as we know it, life requires three basic ingredients: liquid water, a consistent source of gentle warmth, and a small set of basic chemical elements such as carbon, nitrogen, sulfur, and phosphorus. (Liquid water provides hydrogen and oxygen, as well.) If all three of these requirements are found together anywhere on Earth, life is always present; extrapolating to the universe, environments with these requirements may well also harbor life as we know it.

## Why do we **assume** that if we find **conditions on exoplanets** that are **similar** to those on **Earth**, then **life will be found** there?

The study of the universe—perhaps especially the study of life in the universe—often depends on a key assumption called the Copernican Principle. This principle, named after the Polish astronomer who proposed that Earth was not the center of the universe, posits that the same laws of nature hold true everywhere in the universe. Earth is not an exception to that rule; in other words, "we are nothing special." This means that, if life formed on Earth because it had certain characteristics, then any other planet matching those characteristics will have the same chance of eventually supporting life, too. The main question is, which characteristics are the important ones? Scientists think that the keys to life on Earth are liquid water, the right chemicals, and a steady energy supply. It is not certain, though, that these are indeed the correct necessities for life, nor is it certain what kinds of life these conditions could support. What if life on Earth evolved along just one of many possible paths? Astronomers might not even recognize the life on these other planets!

### How can the **Copernican Principle** be applied to **exoplanets** that **do not orbit stars**?

The discovery of exoplanets in very strange orbits—for example, gas giants orbiting their host stars at distances much closer than Mercury is from our Sun—has shown that planets often migrate from the orbits where they formed. That means, in turn, that planets can often be flung out of their planetary systems by gravitational interactions with other migrating planets, as if they were in an interplanetary game of billiards. Although this has not happened in our own solar system for billions of years, the Copernican Principle suggests that, someday, our solar system could undergo such an upheaval as well.

If this planet-flinging scenario proves to be true, then there could be billions of rogue planets flying through interstellar space, free from the gravitational wells of the stars that birthed them. If such a planet had a thick crust and an underground liquid ocean, then tidal or geothermal processes deep in those planetary cores may be warming that ocean, creating a teeming ecosystem that is hurtling unfettered through the galaxy. Could such a planet fly through our own solar system someday? The odds are slim to none, but it is not impossible.

316

Astronomers have discovered young solar systems surrounded by vast amounts of water vapor, as shown in this artist's illustration depicting NGC 1333-IRAS 4B, a system with enough surrounding water to fill Earth's oceans five times over. (*NASA/JPL-Caltech/R. Hurt*)

### What did the *Galileo* spacecraft learn about detecting **life on distant worlds?**

During *Galileo*'s December 1990 flyby of Earth, astronomers trained its instruments and cameras on our own planet. Onboard sensors were able to measure the signatures of life: an atmosphere rich in the two highly reactive gases oxygen and methane. Green light from plants reflected off the surface that covered most of the land surface. Finally, its radar detectors noticed a great deal of radio wave emission emitting within narrow bands of the electromagnetic spectrum—signals too orderly and well-organized to come from lightning, aurorae, or other natural energy bursts—that indicated communication between intelligent beings. Future space probes will use these readings from *Galileo* as a baseline to refer to when searching for extraterrestrial life.

### How has the **discovery of exoplanets affected** the **search** for **extraterrestrial life?**

Since the definitive discoveries of the first exoplanets in the 1990s, hundreds of exoplanets (or "extrasolar planet") have been discovered. Of these detections, dozens have been shown to exist in exoplanetary systems that contain more than one planet orbiting a single star. These discoveries completely changed the way scientists thought about extraterrestrial life. On the one hand, if so many planets exist, and so

317

many planetary systems with multiple planets exist, then surely solar systems like ours exist—and with them, the possibility that they harbor life as we know it. On the other hand, the remarkable variety among planetary systems discovered so far suggests that scientists may have thought too narrowly in the past about environments where life might thrive. Instead of basing the search for life solely on solar systems like our own, or planets like those orbiting the Sun, astronomers are now thinking much more broadly and creatively about ways to find life in the universe.

### Has **any planet** yet been found in a **habitable zone** around a star?

Habitable zones—where the heat from a star would keep water on a planet's surface liquid—exist around most of the stars where exoplanets have been discovered so far. Almost none of their exoplanets, however, are orbiting in those stars' habitable zones. In November 2007, however, a planet was discovered around the star 55 Cancri that appears to be orbiting in the habitable zone. This planet is almost certainly a gas giant planet and not a terrestrial one—its minimum mass is about twice the mass of Neptune. But like the gas giants in our solar system, it may have moons orbiting it that may have rocky or metallic crusts and mantles. Those moons, if they exist, could harbor liquid water. Thus, with its host star providing just the right amount of heat and light, such a moon could harbor life as well.

### What do we need to know about **exoplanets** to **find life** on them some day?

Of all the areas of research in astronomy today, the study of exoplanets is one of the most intriguing. The search for extraterrestrial life, which was previously relegated to the realm of science fiction, has only recently become a viable topic for legitimate scientific study. The combination of searching for exoplanets and for the life forms that might inhabit them is only in its infancy. Scientists are steadily inventing new ways to go about this task, which has no precedent. Ultimately, this new adventure embodies everything that is exciting, inspiring, and just plain fun about astronomy.

In the end, the questions we will have to answer will be ones we have yet to ask.

# Index

Note: (ill.) indicates photos and illustrations.

**319**